Data Generation for Regulatory Agencies: A Collaborative Approach

ACS SYMPOSIUM SERIES **1384**

Data Generation for Regulatory Agencies: A Collaborative Approach

David E. Barnekow, Editor
Corteva Agriscience
Indianapolis, Indiana, United States

Michael E. Krolski, Editor
Bayer CropScience
Chesterfield, Missouri, United States

Sponsored by the
ACS Division of Agrochemicals

American Chemical Society, Washington, DC

Library of Congress Cataloging-in-Publication Data

Names: Barnekow, David E., editor. | Krolski, Michael E., editor.
Title: Data generation for regulatory agencies: a collaborative approach / David E. Barnekow, Michael E. Krolski.
Other titles: ACS symposium series ; 1384.
Description: Washington, DC : American Chemical Society, [2021] | Series: ACS symposium series ; 1384 | Includes bibliographical references and index. | Identifiers: LCCN 2021029055 (print) | LCCN 2021029056 (ebook) | ISBN 9780841298293 (hardcover) | ISBN 9780841298286 (ebook other)
Subjects: LCSH: Data curation. | Data processing. | Data sets. | Administrative agencies.
Classification: LCC QA76.9.D345 D36 2021 (print) | LCC QA76.9.D345 (ebook) | DDC 025--dc23/eng/20211014
LC record available at https://lccn.loc.gov/2021029055
LC ebook record available at https://lccn.loc.gov/2021029056

The paper used in this publication meets the minimum requirements of American National Standard for Information Sciences—Permanence of Paper for Printed Library Materials, ANSI Z39.48n1984.

PRINTED IN THE UNITED STATES OF AMERICA

Foreword

The ACS Symposium Series is an established program that publishes high-quality volumes of thematic manuscripts. For over 40 years, the ACS Symposium Series has been delivering essential research from world leading scientists, including 36 Chemistry Nobel Laureates, to audiences spanning disciplines and applications.

Books are developed from successful symposia sponsored by the ACS or other organizations. Topics span the entirety of chemistry, including applications, basic research, and interdisciplinary reviews.

Before agreeing to publish a book, prospective editors submit a proposal, including a table of contents. The proposal is reviewed for originality, coverage, and interest to the audience. Some manuscripts may be excluded to better focus the book; others may be added to aid comprehensiveness. All chapters are peer reviewed prior to final acceptance or rejection.

As a rule, only original research papers and original review papers are included in the volumes. Verbatim reproductions of previous published papers are not accepted.

ACS Books

Contents

Indexes

Preface

Data generation for regulatory decision-making, particularly for risk assessment purposes, has become both complicated and expensive for the regulated community. Additionally, comparing data generated by a wide variety of submitters, often using different protocols and methodology, makes regulatory decision-making difficult. One approach to addressing these challenges is for the regulated community to collaborate to generate a common dataset. *Data Generation for Regulatory Agencies: A Collaborative Approach* is a compilation of the presentations given in a symposium, *Task Force Data Generation for Risk Assessment,* at the 260th ACS National Meeting, held virtually August 17–20, 2020. The chapters in this volume highlight the formation, operation, and data submission of several Task Forces, along with regulators' use of the submitted data in their evaluation of individual products.

Task Forces for data generation can be divided into three general classes. The first class of Task Force is formed to address a compound-specific or class-specific regulatory question or issue. The purpose of these groups is typically very narrow in scope and the data generated is rarely applicable to a broader agricultural community. Additionally, membership in these Task Forces is limited to a few registrants and the resulting data is highly proprietary. Due to the limited general applicability and associated data confidentiality issues, this class of Task Force is not included in the current volume.

The second class of Task Force is formed to address physical, environmental, and ecotoxicological topics that are common to multiple agricultural environments. In general, they are formed to address the impact of agricultural practices on the environment. The first example of this class of Task Force is described by Dave Johnson in *Spray Drift Task Force (SDTF) – Development of Data and Models for Determining Off-Target Drift of Agricultural Pesticides* (Chapter 3), where a group of registrants came together to address a question that was common to all spray applications. Bernalyn McGaughey describes how a Task Force was formed in response to a specific environmental regulation in *The FIFRA Endangered Species Task Force: Dealing with Unusual Challenges and Multiple Agencies* (Chapter 7). The effect of pesticide applications on beneficial insects is addressed by Joseph Wisk in *Pollinator Research Task Force - Contributing to the Science of Pollinator Risk Assessments for Pesticides* (Chapter 8).

The third class of Task Force is formed to address issues that arise from the exposure of humans to agricultural chemicals. This can be exposure of mixers, loaders, and applicators; reentry agricultural workers; bystanders; or consumers of agricultural commodities. Exposure of mixers, loaders, and applicators is described by David Barnekow in *Agricultural Handler Exposure Task Force –Development of a Generic Exposure Database for Agricultural Workers Performing Mixer/Loader and Applicator Tasks* (Chapter 4) and exposure of reentry workers is covered by Mike Krolski in *Agricultural Reentry Task Force (ARTF) – History, Data Development, and Data Use* (Chapter 5). The potential exposure of bystanders and homeowners is described by Jennifer Thomasen in *Overview of the Outdoor Residential Exposure Task Force (ORETF)* (Chapter 6). A tool for the evaluation of human dietary exposure is described by Jason Johnston in *Use of CARES NG to Conduct a Cumulative Dietary Exposure Assessment for Chlorotriazine Compounds: A Case Study* (Chapter 9).

A case study of the mechanics of running a Task Force and the methodology of data generation and evaluation is described by Eric Bruce in *Agricultural Handler Exposure Task Force, A Technical Overview of Agricultural Handler Exposure Data Development* (Chapter 10). This chapter can serve as an example for the mechanics of running a Task Force successfully through the study planning, data collection, and data submission phases of its life cycle.

Equally important to the generation of data by Task Forces is the use of that data by regulators in their efforts to ensure the safe use of agricultural chemicals. The partnership between the United States Environmental Protection Agency and various Task Forces is described by Jeff Dawson in *Regulatory Perspective On Exposure Task Forces* (Chapter 1) while that between Task Forces and the Pest Management Regulatory Agency of Canada is described by Cristina Vizena in *Regulatory Use of Generic Exposure Data for Pest Control Products: The Canadian Perspective* (Chapter 2).

This compilation is intended to provide the reader with a concise overview of the history and current status of both the scientific and regulatory aspects of multiple Task Forces which were formed to generate data for the regulatory support of pesticide registrations. It is the hope of the editors that it will also be a resource for future collaborative research efforts between regulators and the regulated community. By working together to address challenges confronting the safety of our food supply, agricultural workers and consumers, and the environment, we can continue to find the best solutions.

The editors would like to thank all of the Task Forces represented in this volume and their participating organizations. We thank the contributing authors for sharing their expertise and insights in these collaborative efforts and appreciate the cooperation and guidance of the regulators who helped guide the various Task Forces. Finally, we would like to thank the editorial staff of ACS Books and Technica, particularly Amanda Koenig, Tracey Glazener, and Victoria Balque-Burns, who shepherded this volume through the publication process.

David E. Barnekow
Corteva Agriscience
Indianapolis, Indiana, United States

Michael E. Krolski
Bayer CropScience
Chesterfield, Missouri, United States

Chapter 1

Regulatory Perspective on Exposure Task Forces

Jeffrey L. Dawson*

Science Advisor, Office of Chemical Safety and Pollution Prevention, United States Environmental Protection Agency, 1200 Pennsylvania Avenue, Washington, D.C. 20460, United States

***Email: Dawson.Jeff@EPA.Gov**

Several task forces over the last few decades have generated data for pesticides which are routinely used in Environmental Protection Agency (EPA) risk assessments for evaluating occupational exposures and exposures in the general population. Outputs include unit exposures for pesticide handlers which relate exposure to the amount of pesticide handled for specific conditions (equipment, personal protective equipment use, clothing, etc.) or transfer coefficients which relate exposure to activities which occur in previously treated areas (vegetable harvest, tree pruning, etc.). Data call-in authority under Federal Insecticide, Fungicide, and Rodenticide Act (FIFRA), sparked the development of these groups. EPA and other regulatory agencies provided input for each task force during their life cycles on several topics: program scope, study development, statistical design, standard procedures development, peer review, regulatory compliance, and data analysis. The resulting data from these efforts are used broadly on a global basis and they represent one of the largest dermal sampling databases in existence across all chemical space.

The task force approach has many benefits. It provides a forum for impacted stakeholders to coalesce to develop information that can be used broadly for regulatory purposes in a uniform manner. This allows for more consistent regulatory decision-making. It also allows for collaborative input from all involved including regulators and impacted stakeholders through all phases of the data development process which makes it more efficient and well documented. The EPA completes hundreds of risk assessment actions for pesticides each year. Task force data play a significant role in each of these which involve non-dietary exposures.

Introduction

Exposure task forces, groups of companies formed to address a common purpose under the Federal Insecticide, Fungicide, and Rodenticide Act (FIFRA), have had a profound impact on how regulatory agencies approach risk assessment for non-dietary exposures from pesticides. Collaborative efforts between multiple regulatory agencies and task force memberships have been successful. This chapter explores the regulatory perspective on these groups by examining how regulations were developed without task forces, their histories and development, the impact of the 2006 revisions to the human studies rule in the United States, current ongoing efforts, the products which were developed, and the programmatic benefits. The remainder of this book will address the details associated with various task force efforts, regulatory perspectives on their use from other agencies, and efficiencies for the regulated community.

Before Task Forces

The founding of the Environmental Protection Agency (EPA), statutory changes such as the 1978 amendments to the FIFRA, and societal desires focused on environmental improvements created a nexus focused on the evaluation of risks for those occupationally exposed to pesticides. The EPA responded through staffing and structural changes to accommodate these issues. In 1980 the National Research Council (*1*) that was tasked to consider EPA's efforts in this area concluded:

> *"In estimating exposure, OPP [Office of Pesticide Programs] is constantly hampered by lack of adequate data and is forced to resort to indirect and inaccurate methods in its efforts to make plausible estimates."*

> [Note: OPP is the program in the EPA responsible for pesticide regulation.]

Following an EPA administrator directive in 1978 (*2*), one of the first steps to address these regulatory difficulties was the issuance of data collection guidelines (*3*), that were based on sampling methods available in the scientific literature (*4*). During the period of 1980 to 1984, the EPA began systematic evaluation of exposure data and field work poisoning incidents. These efforts and the evaluation of the first sets of disparate exposure data based on monitoring protocols lead to the development of the first generation of exposure and risk assessments. These assessments were focused on workers who mix, load, and apply pesticides ("handlers") and those involved in hand labor activities in production agriculture.

Production of exposure and risk assessments during this period was highly resource intensive and often resulted in inconsistent findings due to data limitations inherent with the use of information which was quite varied based on the data associated with each chemical. Preparation of fully comprehensive individual assessments for all chemicals would have required chemical-specific data first be available and then reviewed for use in the assessment process. Given there are approximately 600 conventional (*5*) pesticide active ingredients, each typically having many associated exposure scenarios for both those who handle such materials and who work in previously treated areas, this chemical-specific approach could have resulted in requiring many thousands of exposure studies. Based on how a particular pesticide can be used, scenarios are defined by many physical factors such as formulation type, equipment used in application processes, protective measures used, crop type, and the types of hand labor activities required for production. For example, a liquid product in typical packaging would always have a scenario for open mixing-liquids where the exposure estimates reflected at a minimum label prescribed clothing and/or protective measures (e.g., glove use). Assuming, for example, 25 scenarios associated with each of approximately 600 chemicals, results in 15,000 studies being needed. If 25 test subjects were monitored in each of these studies this would equate to 375,000 subjects being monitored. If current costs are used (~$75,000 per monitored person per day) the total costs of developing such data would be approximately $28 billion. This does not include the extensive costs to the EPA related to data review and analysis and exposure assessment considerations. These resource predicaments plagued both the regulated industry and the EPA in this period. Also, during this period all impacted parties were struggling with consistent application of principles related to data generation, data analysis and utility of data in the exposure assessment process, as evidenced in efforts such as the Rejection Rate Analysis completed by the EPA (*6*). All of these factors contributed significantly to equitable scientific consideration of chemicals in the regulatory framework managed by the EPA. This led to conceptualization of more efficient approaches such as task forces for potentially conducting exposure assessments in a more systematic and efficient manner.

Task Force Origins and Underlying Concepts

There is a statutory context for the concept of task forces operating to develop data needed for pesticide registration purposes as outlined in the FIFRA. FIFRA § 3(c)(2)(B) permits two or more registrants to cooperate and develop data required for registration. Task forces, therefore, may share costs to purchase existing proprietary data or they may share costs to develop new data. Costs shared by task forces include the direct cost of developing data (e.g., research personnel, laboratory analysis, etc.), cost of management, and overhead related to submission of data. Task forces operate under legally binding agreements defining their purpose and they often form to address specific Data Call-In (DCI) requirements from EPA.

Given the chaotic environment in which both regulatory agencies and the regulated community were operating at the time (early to mid-1980s) because of the inconsistencies inherent in the exposure assessment process and the sizable resources needed to meet statutory requirements, it was clear there was a need for alternative approaches. A seminal event was a publication where development of a "generic" database was proposed for assessing exposures for pesticide handlers (*7, 8*).

This ultimately resulted in the development of the Pesticide Handlers Exposure Database (PHED) (*9*). PHED was based on a data waiver agreement by registrants who provided their data for no compensation so long as there were no study-identifying attributes in the database. PHED, released in 1992 then updated in 1995, represented a gigantic leap forward in the discipline of exposure assessment not only from a scientific but also from policy and administrative perspectives as well. PHED contains data for approximately 1700 monitoring units (MUs) for people loading and/or applying pesticides, including flagging for aerial operations. These events represent exposure monitoring of one person during one work cycle. For each MU there are many data points which document the application and climatic conditions of the event. Also, data were collected regarding dermal exposure levels across various body parts and inhalation exposures along with the corresponding quality control information such as analytical recoveries. There are over 100 studies represented in the data within PHED and they are disparate in design. This makes certain statistical use of the data impossible because of inconsistencies in the data. The software also allowed users to segment data based on parameters which relate to the physical attributes of each monitoring event (e.g., equipment used, clothing worn, weather conditions). There were significant negative issues related to how the data were shared among the data donors to the software associated with intellectual property rights associated with refined pesticide handling technologies. Because of these issues, data were no longer input into the software and the last update was 1995. PHED was used to develop "unit exposures", which are a key metric used in pesticide handler exposure assessment. Unit exposures relate exposure levels to the amounts of pesticide used specific to the exposure scenario (i.e., product, equipment, and level of personal protection employed). PHED was used to develop a surrogate guide which is a reference document that provides a standard, central source for unit exposures used in assessments (*10*). Creation of the surrogate guide allowed for more efficient and consistent development of exposure estimates in the risk assessment process. Equally important, this resulted in more systematic and equitable treatment of the regulated community and consistency in regulatory decision-making.

The EPA is responsible for evaluating risks across the entire life cycle of pesticides. This means potential exposures resulting from all allowable use patterns must be considered. These exposures include handlers involved in applications and those who may be exposed when working in previously treated areas performing activities such as hand harvesting fruits and vegetables. This also pertains to other environments such as outdoor and indoor residential settings where pest control or other treatments have occurred. There was a lack of data available for these other environments as well and, given the experience with data sharing associated with PHED, the EPA opted to seek data through more formal processes prescribed under FIFRA. In response, the EPA issued two DCIs using its authority under FIFRA (*11*). These DCIs focused on the major issues of the time related to exposure assessment: agricultural reentry exposures (e.g., due to hand labor such as fruit harvest) and outdoor residential exposures (e.g., due to lawn and garden applications of pesticides). Other mechanisms, such as Pesticide Registration Notices (PRN), were also employed to publicly identify data needs for other types of activities, such as the need for additional pesticide handler exposure data, the need for exposure data in indoor residential environments, and the need for use and usage data especially in the consumer marketplace (*12*).

Out of these efforts, some basic conceptual premises were developed which remain to this day. The conceptual premises have also been tested to ensure their rigor as described in other chapters of this publication. For handlers, exposures are not believed to be a function of the active ingredient, but rather are considered to be based on the conditions an individual would experience. Determinants of handler exposures include the form of the product used (solid or liquid), the type of solid product used (powder or granule), the packaging of the product (open pour or closed system), the application equipment used (ground, aerial, or handheld equipment), the use of personal protective equipment (additional clothing, gloves, respirators), and the sophistication of the application system dictated by the use of tractor cabs, closed loading/transfer systems and other types of engineering controls. Focusing on these determinants leads to the premise of surrogacy or genericness. If applications of two chemicals (A and B) occurred at identical rates using the same equipment and same levels of personal protection, the resulting exposures would be expected to be equivalent regardless of the identity of chemical (A or B).

Similarly, in reentry situations, exposures are believed to be generic and can be described as a function of the type of crop or area where an individual is present, defined by the types of crops, the specific agronomic activities used for crop production, and the specific growth stage when activities occur. The nature of the activity also plays a key role in the levels of anticipated exposures reported as "transfer coefficients" (TCs). TCs are metrics which define exposure rates for specific crop/activity combinations and they vary based on the level of effort and crop type/maturity. For example, hand harvesting crops is typically a fast and efficient process that requires substantial contact with treated foliage by workers given that they retrieve the desired edible component of the crop. TCs for these types of activities are relatively high compared to other activities conducted within the same crop that results in less foliar contact, such as using hand tools to trim or hoe. The genericness concept applies to reentry situations as well. Exposures would be similar for an individual exposed to the same level of foliar residues of two different pesticides if all other conditions were equal – doing the same activity in the same field where residues of chemical A or B were the same. This surrogacy concept assumes chemicals are indicators of exposure, not a component or predictor of exposure. As a term of art, the moniker *surrogate chemical* is often utilized to describe this issue. Any impacts of individual chemical differences are related to, and addressed via, analytical treatments of the data (e.g., adjustments based on analytical stability, recovery, etc.).

In summary, the use of a task force approach eliminates broad use of chemical-specific data and approaches which are inherently inconsistent and resource intensive. The assumption of genericness is critical. It allows for development of universally applicable databases that are completed effectively using the task force construct. Development of data by task forces reduces resource needs, alleviates concerns over inconsistency, and enhances rigor in the data used for regulatory purposes.

Specific Task Forces

Much of the remainder of this book will address the specifics of the major generic data task force efforts which developed over approximately the last 30 years. They have been focused on generation of data which can be systematically employed by regulatory agencies and the regulated community to ensure transparent and equal treatment in the pesticide regulatory process. These efforts also enhance transparency for all stakeholders such as agricultural producers and advocacy groups. Finally, the data have a rigorous foundation because the data development process actively involved an array of impacted parties whose considerations and concerns were incorporated and addressed throughout the process. For example, the methods for identifying sampling locations and workers in the U.S. evolved over time to include elements of random selection to ensure more statistically robust research designs (covered in more detail later in this publication).

Major task force efforts have been/are focused on the development of surrogate data to be used broadly across the regulated community. Specific key groups include:

- Pesticide Handlers Exposure Database (PHED),
- Agricultural Reentry Task Force (ARTF),
- Outdoor Residential Exposure Task Force (ORETF),
- Non-Dietary Exposure Task Force (NDETF),
- Antimicrobial Exposure Assessment Task Forces I and II (AEATF),
- Agricultural Handlers Exposure Task Force (AHETF),
- Residential Exposure Joint Venture I & II (REJV), and
- Pet Care Products Task Force (PCPTF).

[Note: While PHED was not a formal task force which was a root cause of the data sharing issues noted earlier, it is mentioned here because it contained many of the key attributes of current task force efforts and was the prototype in many ways for how they developed.]

The major generic data generating task forces are noted in Table 1. The table provides information for each relative to their scope.

Table 1. Scope of Major Task Force Efforts

Scenario	*Task Force*	*Populations/Settings To Which Data Apply*
Handler Exposure	PHED, ORETF, AEATF	Indoor and Outdoor Residential/General Population
	PHED, ORETF, AHETF, AEATF	Indoor and Outdoor Occupational
Reentry Exposure	ARTF, ORETF	Outdoor Residential and General Population
	NDETF, PCPTF	Indoor Residential and General Population Indoor and Outdoor Pets
	ARTF	Occupational Outdoor and Greenhouses
Use Parameters	REJV	Indoor and Outdoor Residential

Impact of EPA's Human Studies Rule

Perhaps no aspect of how human exposure research is completed to satisfy FIFRA data requirements has evolved so significantly as the protection of human subjects. The rule *Protections for Subjects in Human Research* (40CFR26), was promulgated in early 2006 (*13*). It extended the ethical protections in the "Common Rule (*14*)" to other human research involving intentional exposure of non-pregnant or non-nursing adults intended for submission to EPA under the pesticide laws.

The rule also banned all third-party intentional dosing research on pesticides involving intentional exposure of pregnant or nursing women and children intended for submission to EPA. Additionally, it required that EPA neither conduct nor support any intentional dosing studies which involve pregnant or nursing women or children for all substances that EPA regulates. The rule also provided ethical protections to adult subjects who volunteer to participate in intentional exposure studies for pesticides.

The Human Studies Review Board (HSRB) (*15*) is a body governed by the Federal Advisory Committee Act (FACA) was also established based on requirements of the rule. HSRB ensures third-party proposals for new research and completed intentional dosing research on which EPA may rely under the pesticide laws is the subject to independent expert peer review.

In addition, the rule adapts regulations of the Department of Health and Human Services providing additional protections beyond those of the Common Rule to pregnant women and children as subjects in EPA observational research that does not involve intentional exposure to any substance.

Amendments to the rule adopted in 2013 broadened its scope and strengthened the standards for research considered in EPA decisions. The amendments also clarified the approach used for science and ethics reviews completed by the HSRB and disallowed participation in research for those who cannot offer informed consent for themselves.

In practical terms, the changes to the human studies rule most significantly impacted two task forces – AHETF and AEATF. These two task forces were active after the rule changes and thus the rule significantly impacted their programs. Because the research by these task forces was considered subject to the rule, HSRB reviewed both scientific and ethical elements of each program. As a result of these reviews each task force incorporated much more rigorous standard operating procedures designed to protect human subjects. They focused on issues such as how informed consent was obtained, how subject's health was protected during the research, and how information regarding subject participation in the research could be utilized. Also, aspects of their study designs were revised to enhance the scientific rigor of this type of research. Most significantly, much more detail was added to describe how subjects and settings are selected for monitoring purposes to achieve the most statistically representative sampling strategy possible (e.g., random selection of names from call lists to identify study participants). This impacted the nature of the recruiting process which became much more formal and substantive based on predetermined design strategies to ensure broad, applicable ranges of exposure conditions were identified and that subjects within those exposure conditions were selected based on as many random elements as possible.

These requirements had significant initial, front-end costs, but once the detailed requirements and approaches were developed in a manner satisfactory to the HSRB, the research continued at a predictable pace. These requirements also added a level of rigor to the results to make them much more incontrovertible, especially given a major previous criticism of such task force programs that monitoring was not conducted using representative populations. The current approaches reflect the requirements of the rule, as interpreted by the HSRB, alleviate much of these criticisms.

Key Task Force Outputs and EPA Products

EPA exposure assessments as noted earlier in Table 1 consider a variety of populations and settings. Comprehensive assessments typically evaluate handler and postapplication (or reentry) exposures in both occupational and residential (general population) settings.

Depending on the uses associated with a pesticide, occupational handler assessments can consider more than 1000 scenarios where each scenario is based on unique attributes associated with a particular use. EPA's handler scenarios are defined in part by pesticide product labels which instruct users on where and how to apply the product and specify the clothing and protective equipment required for use. Other factors include the types of equipment used in applications and the parameters of applications such as rates, frequency, crops/targets, and crop growth stages. Exposures are calculated by determining how much material can be applied in a typical day generally based on the product of application rate and the area treated. The area which can be treated is a function of the crop or type of target to be treated and the equipment used. For example, a tractor-mounted sprayer would treat more area than a handheld piece of equipment. Once the amount handled in a day is determined, it is multiplied by an exposure metric referred to as a "unit exposure". Unit exposure estimates are the outputs of pesticide handler exposure task forces, and they typically relate exposure to the amount of material handled specific to the exposure condition. The values developed based on these data by EPA are publicly available in the "Occupational Pesticide Handler Unit Exposure Surrogate Reference Table (Rev. March 2020)" along with a Microsoft Excel-based calculator and background information on their derivation (*16*).

Occupational postapplication (or reentry) exposure assessments can consider as many as 4500 or so scenarios depending upon the uses associated with a pesticide. Like with handler exposure assessments, pesticide product labels instruct users on where and how to apply the product. This in turn defines the postapplication assessment scenarios. Key information includes the crop, application rates, and the timing of application relative to the stage of crop growth. Additionally, the agronomic practices required to produce crops such as weeding, scouting, pruning, thinning, and hand harvesting are critical in defining exposure scenarios. Metrics derived from ARTF exposure monitoring data referred to as TCs are used to calculate dermal exposures because they represent the amount of contact with previously treated crops for a particular crop and activity combination. TCs are multiplied by the amount of transferable residue on the contacted surfaces (e.g., crop foliage) and by the duration of exposure associated with the activity (e.g., for occupational scenarios, typically an 8-hour workday). Background information on the ARTF data is publicly available along with all the crop/activity-specific TCs in the "Science Advisory Council For Exposure (ExpoSAC) Policy 3 – Revised January 2017 (*17*)." There is also a Microsoft Excel-based calculator that integrates the information along with the algorithms necessary to complete an analysis.

Residential handler and postapplication exposures are considered for all types of scenarios associated with uses of pesticides where the general population can be exposed. These settings can include lawn/turf treatments, indoor pest control uses, garden plant treatments, pet treatments, and uses on clothing. There are other areas where the general population can be exposed such as in public areas like parks, athletic fields, golf courses, and institutional settings such as schools and daycares. The methods used for these types of analyses are contained in the publicly available "Standard Operating Procedures for Residential Pesticide Exposure Assessment" (SOPs) (*18*). The SOPs contain information such as unit exposures and TCs generated by task forces. These are used to evaluate both residential handler and postapplication exposures. Supporting information and Excel-based calculators are also available.

Programmatic Benefits to the EPA

The EPA tools described earlier inherently rely on the use of task force data from many sources. They are central to the core regulatory processes completed by the EPA on a routine basis which equates to several hundred risk analyses per year, covering all types of products and decisions. Task force data generation processes and extensive levels of regulatory input and peer review ensure that the information is of high quality and fit for the purpose of providing a basis for exposure assessment. Use of task force data also provides a basis for consistent approaches and findings in the risk assessment process. This increases efficiency and improves consistency for the regulated community. It also enhances equity in regulatory decision-making processes. Finally, it provides transparency for all other stakeholders, not just the pesticide industry.

Within the regulated community, the program/data development process used by task forces involves all member companies and impacted regulatory agencies. This tends to lead to more thoughtful and consistent research design. It also leads to more consistent "buy-in" and less acrimony during regulatory processes. Historically, there has also been broad international acceptance of data developed by task forces and the surrogate approaches to using the data which is key for global harmonization purposes. Resource savings are a significant downstream result from reducing global inconsistency.

Conclusions

The data developed by task forces are integral to the modern regulatory landscape. The products of these data development efforts, such as unit exposures for pesticide handlers and TCs for postapplication exposures are core elements of regulatory risk assessment programs. Task forces have many benefits including large scale, rigorous research planning including consideration of statistical designs. Such approaches address key criticisms related to the representativeness of the data and how it relates to actual agricultural practices.

There are typically many participants in task forces so they also tend to have the financial resources and technical levels of expertise to rigorously address their scope. Data development by task forces is also mutually beneficial to the regulated community and applicable regulatory agencies. It allows for collaborative processes during each phase of development including protocol development, development of standard operating procedures, addressing ethical concerns of involving humans in research, data analysis and so on. Collaborative inputs on each phase of data development reduces unanticipated surprises related to the use of data. This allows for more informed and rigorous regulatory decision-making.

References

1. National Research Council. *Regulating Pesticides*; The National Academies Press: Washington, D.C., 1980; https://doi.org/10.17226/54.
2. *Report on S.1678 Federal Pesticide Act of September 19, 1978*. H10119. Congressional Record (1978).
3. Pesticide Assessment Guidelines, Subdivision K Exposure: Reentry Protection 1984; U.S. EPA, Washington, D.C. [NTIS Document PB85-120962] and Pesticide Assessment Guidelines, Subdivision U Applicator Exposure Monitoring; U.S. EPA 1986, U.S. EPA, Washington, D.C. [NTIS Document PB87-133286].
4. Durham, W. F.; Wolfe, H. R. Measurement of the exposure of workers to pesticides. *Bull. WHO* **1962**, *26*, 75–91 and Field Surveys of Exposure to Pesticides, Standard Protocol, Technical Monograph No. 7, Pesticide Development and Safe Use Unit, Division of Vector Biology and Control, 1982, WHO HQ, Geneva, Switzerland.
5. Conventional pesticides in this case do not include biopesticides or antimicrobial chemicals. For this purpose, they can be thought of as chemically based compounds such as carbamate insecticides or phenoxy acid herbicides.
6. *Pesticide Reregistration Rejection Rate Analysis – Occupational and Residential Exposure*; U.S. EPA, 1993, EPA/OPPTS [Document ID EPA 738-R-93-008].
7. Hackathorn, D. R.; Eberhart, D. C. Data-Base Proposal For Use In Predicting Mixer-Loader-Applicator Exposure. In *Dermal Exposure Related to Pesticide Use*. Honeycutt, R. C., Zweig, G., Ragsdale, N. N., Eds.; ACS Symposium Series 273; American Chemical Society: Washington, DC, 1985; pp 341–355. ISBN 9780841208988.
8. "Generic" as well as the term "surrogate" are applied to denote external exposure to the skin or levels in the breathing zone which are typically how these types of exposures are reported. These values do not account for processes such as dermal absorption which would make the values chemical-specific and therefore discount the genericness concept.
9. U.S. EPA Occupational Pesticide Handler Exposure Data – Pesticide Handler Exposure Database (PHED). https://www.epa.gov/pesticide-science-and-assessing-pesticide-risks/occupational-pesticide-handler-exposure-data#phed (Accessed 2021-02-25).
10. U.S. EPA Occupational Pesticide Handler Unit Exposure Surrogate Reference Table. https://www.epa.gov/sites/production/files/2020-03/documents/opp-hed-pesticide-handler-surrogate-unit-exposure-table-march-2020.pdf (Accessed 2021-02-25)
11. Data Call In Notice For Registrants With Products With Potential For Reentry Exposure, October 1985, U.S. EPA, Washington, D.C.
12. U.S. EPA (09/20/07) Pesticide Registration (PR) Notice 2007-3: The Agricultural Handler Exposure Task Force L.L.C.
13. U.S. EPA Protections For Human Subjects in Research with Pesticides. https://www.epa.gov/pesticide-advisory-committees-and-regulatory-partners/protections-subjects-human-research-pesticides#:~:text=Under%20EPA's%20rule%20titled%20Protections,will%20neither%20conduct%20nor%20support (Accessed 2021-02-25)

14. U.S. Department of Health & Human Services Federal Policy For Human Subjects. https://www.hhs.gov/ohrp/regulations-and-policy/regulations/common-rule/index.html (Accessed 2021-02-25)
15. U.S. EPA Human Studies Review Board. https://www.epa.gov/osa/human-studies-review-board (Accessed 2021-02-25)
16. U.S. EPA Pesticide Handler Exposure Data Home Page. https://www.epa.gov/pesticide-science-and-assessing-pesticide-risks/occupational-pesticide-handler-exposure-data (Accessed 2021-02-25)
17. U.S. EPA Occupational Pesticide Post-application Exposure Data. https://www.epa.gov/pesticide-science-and-assessing-pesticide-risks/occupational-pesticide-post-application-exposure (Accessed 2021-02-25)
18. U.S. EPA Standard Operating Procedures for Residential Pesticide Exposure Assessment. https://www.epa.gov/pesticide-science-and-assessing-pesticide-risks/standard-operating-procedures-residential-pesticide (Accessed 2021-02-25)

Chapter 2

Regulatory Use of Generic Exposure Data for Pest Control Products: The Canadian Perspective

Cristina Vizena,* Trevor Satchwill, Shairoz Ramji, and Isabelle Pilote

Health Canada, Pest Management Regulatory Agency, Ottawa, Ontario, Canada K1A 0K9

***Email: cristina.vizena@canada.ca**

Health Canada's Pest Management Regulatory Agency (PMRA) is responsible for the federal regulation of pesticides in Canada. According to the *Pest Control Products Act*, companies intending to sell pest control products in Canada must provide data demonstrating the safety of these products to both human health and the environment, and also that these products have value when used according to label directions. Only pest control products that have acceptable risk are registered for use in Canada. The PMRA data requirements for human health risk assessments include toxicology studies to characterize the hazard profile of a pesticide and studies to assess the degree and nature of exposure to specific human populations, such as workers and children. As the science of risk assessment has evolved, so too have the regulatory requirements for exposure studies. While the protocols of chemical-specific exposure studies are typically required to adhere to established guidelines, variations between study protocols have limited the ability to use the data in a more generic fashion. Furthermore, differing policies on the requirements and use of these studies have sometimes led to differences in interpretation between regulatory agencies. As such, experts, working in conjunction with regulatory agencies and researchers, have developed the scientific methodology that allows exposure studies conducted with one pesticide to be used for the assessment of many pesticide active ingredients in a generic manner, such as handler exposure studies. The availability of scientifically robust studies that can be used in a generic manner has enabled the development and standardization of protocols and the generation of human exposure data that is applicable to a wide variety of pesticide scenarios and submissions. Much of the generic exposure data has been developed by the exposure task forces who have incorporated regulatory input throughout the process in a collaborative and scientific manner. Overall, this has resulted in a large, comprehensive, and modern collection of data, and has been an efficient mechanism for developing data required for pesticide submissions. It has also allowed pesticide regulatory authorities, such as the PMRA, to use these studies, when applicable, in lieu of chemical-specific data.

Health Canada's Pest Management Regulatory Agency (PMRA) is responsible for the federal regulation of pesticides in Canada under the authority of the *Pest Control Products Act* (PCPA). Under the PCPA, companies who wish to sell pesticides in Canada must submit data demonstrating the safety of these products in terms of human health and the environment, and also that these products have value when used according to label directions.

Health Canada's scientists review scientific studies submitted by applicants in support of the registration of pesticides, along with information in the public domain, such as academic research or information gathered through post-market surveillance. These include toxicology studies to assess the hazard of a chemical and exposure studies to assess the degree and nature of exposure to specific human populations. The risk assessments consider direct and indirect to exposures to the general population as well as vulnerable populations such as pregnant women, infants, children, women, and seniors. Only pesticides that have acceptable risk (can be used safely) and value, when used according to label directions, are registered for use and can be sold in Canada.

The risk to human health is determined based on the hazard of the pesticide and an assessment of exposure. Exposure assessments are conducted using data from exposure studies, which characterize exposure for persons handling pesticides during mixing, loading, and application, and when contacting treated foliage and surfaces when entering treated areas.

Evolution: From Chemical-Specific Studies to Generic Use of Exposure Data

Chemical-specific exposure studies for various crop and use scenarios can be used to assess pesticide exposure. These studies are designed according to established guidelines and can be specific to the precise manner in which the pesticide is intended to be used. This, however, results in the need for the applicant to generate a multitude of studies in order to characterize and estimate the extent of human exposure for each specific chemical. In addition, the specificity of these studies leads to limitations in terms of the ability to use the data (e.g., in the assessment of another chemical). Furthermore, differing policies on the requirements and use of these studies in some cases can lead to differences of interpretation between regulatory agencies.

In the 1980s, the concept of developing generic pesticide exposure data was explored. Through a comparison of several studies conducted with different chemicals, but using the same pesticide application techniques, it was established that the dominant factor determining human exposure was the physical activity of application (*1*). It was further revealed by Hackathorn and Eberhart (*2*) that exposure is not determined by the properties of the chemical in a formulated product, but rather by physical factors and activities involved in handling the product. This led to the concept that data generated for one chemical could be used to represent exposure to other chemicals used in a similar manner. In these cases, the data and their utility are considered 'generic' in nature. This broadly accepted approach for using exposure data in a generic manner to conduct exposure assessments is supported by regulatory authorities throughout North America and Europe.

The concept that handler exposure is more dependent on the physical activity of mixing, loading, and applying pesticides rather than the nature of the chemical led to a proposal to develop a database of generic agricultural mixer, loader, and applicator exposure studies. This eventually became the Pesticide Handlers Exposure Database (PHED) and was based on a literature review of existing published and unpublished studies. It was developed through a collaborative effort by Health Canada (PMRA), the United States Environmental Protection Agency (US EPA), California Department of Pesticide Regulation (CalDPR) and industry experts. Using the PHED software, the Canadian Surrogate Exposure PHED tables were generated for four broad exposure scenarios:

- Mixer/loader
    - Contains four mix/load formulation files (dry flowables, granules, liquids, wettable powders)
    - Includes both open and closed mix/load scenarios for each formulation type
- Applicator
    - Contains nine applicator files (aerial, aerosol, airblast, groundboom, granular bait dispersed by hand, paintbrush, airless sprayer, right-of-way sprayer, and solid broadcast sprayer)
- Mixer/loader/applicator
    - Contains four mixer/loader/applicator files (belly grinder, low pressure handwand [liquid and wettable powder], backpack, high pressure handwand)
- Flagger

While PHED represented an advancement for exposure assessment, it still had limitations. For example, there were inconsistent study designs (differing measurements and methods), inconsistent data quality, limited data for some exposure scenarios, inadequate analytical limits/frequent non-detects, inadequate number of body parts measured, and some scenarios not reflecting contemporary practices. Despite these limitations, PHED pioneered a path forward for other generic databases with an emphasis on the development of high-quality generic exposure data through a collaborative process between PMRA, regulatory partners, and industry experts (i.e., exposure task forces).

Since the development of PHED, and in recognition of its limitations, there has been a further collaborative effort between exposure task forces and regulatory agencies to develop additional exposure studies for use in a generic manner. Since exposure is scenario-specific but generic in terms of typical chemical properties, the most efficient means of generating exposure data was for individual pesticide registrants to pool their resources for the development of a generic exposure database applicable to virtually all of their products. The formation of exposure task forces resulted in a standardization of protocols and the generation of human exposure data that could be applied to a wide variety of pesticide scenarios resulting in a large, comprehensive, and modern database. This approach also benefits the regulatory agencies by providing them with a single, comprehensive database of high-quality data applicable to most active ingredients rather than relying on a multitude of individual studies of varying quality. The availability of high-quality generic exposure data facilitates a timely, efficient, and equitable pesticide registration process. While PHED is specific to chemical handler assessments, the concept of generic exposure data applies to other scenarios such as post-application and residential exposures. As such, generic exposure data have been developed for additional scenarios.

Some examples of exposure task forces include:

- Agricultural Handlers Exposure Task Force (AHETF)
    - This task force generated a generic database of exposure scenarios for agricultural workers during mixing, loading, and application activities.

  - Covers more exposure scenarios than PHED.
  - High-quality data that meets modern regulatory standards.
  - All mixer, loader, applicator studies are conducted in accordance with USEPA OPPTS Guideline Series 875 Part A.

- Outdoor Residential Exposure Task Force (ORETF)

  - This task force generated a generic database of exposure data applicable to homeowners who mix and apply pesticides to their yards, professional lawn care operators who mix and apply pesticides to residential turf areas, and to individuals who enter a turf area following pesticide application.

- Non-Dietary Exposure Task Force (NDETF)

  - This task force generated a database of human and human environment exposure levels associated with residential and some occupational uses of natural pyrethrins, synthetic pyrethroids and associated synergist piperonyl butoxide and N-octyl bicycloheptene dicarboximide (MGK 264).

- Agricultural Re-entry Task Force (ARTF)

  - This task force generated a database of generic agricultural transfer coefficients that are applicable to post-application crop/activity scenarios.
  - The crop and activity specific transfer coefficients, along with chemical-specific dislodgeable foliar residue/turf transferable residue data, are used to conduct the exposure assessment for workers entering treated areas.

How PMRA Uses Generic Exposure Data

The exposure task forces interact with Canada's PMRA and United States counterparts (US EPA and CalDPR) to ensure the data generated is representative of the requirements for pesticide registration both for new pesticide products as well as for active ingredients undergoing re-evaluation. In addition, joint regulatory participation has facilitated the alignment of policies on data requirements and study interpretation. For each study generated by the exposure task forces, comprehensive study reports and full datasets are made available to PMRA for analysis. The studies used for each scenario are reviewed and rigorously compared to accepted regulatory study guidelines. Standard exposure estimates are generated from these data to be used in exposure assessments. To promote transparency of the regulatory process, these values may be published in regulatory documents or provided upon request.

Mixing, loading, and application exposure data generated by the task forces is often representative of workers wearing a single layer of clothing, which may include chemical-resistant gloves depending on the scenario. Inhalation exposure data are representative of workers who are not wearing respiratory protection. In order to estimate exposure for workers wearing higher levels of personal protective equipment (PPE), such as coveralls or a respirator, PMRA incorporates protection factors into the exposure assessment. Protection factors are specific to each type of PPE and are based on how effective it is at reducing exposure.

Commercial Mixer, Loader, and Applicator Scenarios:

For handler scenarios, exposure studies, including those generated by task forces, are used to determine 'unit exposures'. Unit exposure are the average exposure for workers conducting the activity, normalized by the amount of active ingredient that is handled by workers in that study. For workers mixing, loading, and applying pesticides, PMRA typically uses unit exposures to calculate dermal and inhalation exposure using the following equation:

$$\textbf{Exposure (mg/kg bw/day)} = \frac{\textbf{UE x Rate x DA x ATPD x CF}}{\textbf{BW}}$$

Where:

Exposure: Dermal or inhalation normalized by body weight. Values are combined with the hazard of the pesticide to calculate risk.

UE: Re Unit Exposure (μg/kg of active ingredient). These are generated from exposure studies, including studies generated by the exposure task force (AHETF, ORETF, etc.).

Rate: Application rate of active ingredient (kg/ha). The amount of pesticide applied to an area.

DA: Dermal Absorption (%). The amount of active ingredient that is estimated to be absorbed through the skin and is applicable to the dermal route of exposure. For the inhalation route of exposure, absorption is assumed to be 100%.

ATPD: Area Treated per Day (ha/day). The area that the pesticide is applied to on the day of application.

CF: Conversion Factor (mg/1000μg).

BW: Body Weight (kg). Average body weight of males and females.

Commercial Post-Application Exposure Scenarios:

For post-application scenarios, exposure studies, including those generated by task forces, are used to determine 'transfer coefficients'. Transfer coefficients are the average exposure for workers conducting the activity normalized by the amount of residues that are transferred or dislodged from the treated surface. For workers entering treated surfaces, including foliage or turf, PMRA typically calculates dermal post-application exposure using the following equation:

$$\textbf{Exposure (mg/kg bw/day)} = \frac{\textbf{TC x DFR/TTR x DA x ET x CF}}{\textbf{BW}}$$

Where:

Exposure: Dermal or inhalation normalized by body weight. Values are combined with the hazard of the pesticide to calculate risk.

TC: Transfer coefficient (cm^2/hr). The amount of treated foliage that a worker contacts while performing a specific activity in a given period.

DFR/TTR: Dislodgeable Foliar Residue/Turf Transferable Residue (μg/cm^2). The potential residue that can be dislodged from a surface, such as the leaves of a plant or the top of turf.

DA: Dermal Absorption (%). The amount of active ingredient that is estimated to be absorbed through the skin and is applicable to the dermal route of exposure.

ET: Exposure Time (hr). This is based on a typical 8 hour workday.

CF: Conversion Factor (mg/1000μg).

BW: Body Weight (kg). Average body weight of males and females.

Residential Scenarios:

The availability of generic data has also resulted in the PMRA's adoption of policies outlined in the 2012 US EPA Residential Standard Operating Procedures (SOPs) for calculating residential handler and residential post-application exposure. These SOPs provide specific models to assess the different exposure scenarios. The commercial handler and post-application exposure models described previously are similar to those used to assess residential scenarios. Exposure data is representative of persons wearing shorts, short-sleeved shirts, and no gloves or respiratory protection.

Benefits of Using Generic Exposure Data

The PMRA uses the generic data generated through the exposure task forces and other data generators on a routine basis as they are applicable to many pesticide products. Chemical-specific studies continue to be acceptable for regulatory purposes as an alternative to generic data, or for refined assessments. In some cases, higher tiered chemical-specific exposure studies are helpful to further inform the risk assessment, for example, biomonitoring studies.

There are many benefits to using generic exposure data. Data generated through standardized study protocols greatly improves the reliability and confidence in the information and ensures that the variability in specific exposure scenarios is captured. The collaborative and scientific approach used by all experts involved, which includes attaining regulatory input throughout the process, has resulted in a large, comprehensive, and modern collection of data, and has been an efficient mechanism for developing data required for pesticide regulation. Overall, the regulatory experience is that the use of generic exposure data has provided greater confidence and improved quality and consistency in risk assessment. The PMRA's participation in these task forces has also increased PMRA's ability to identify and address critical gaps in worker exposure data, a greater ability to inform regulatory policies and decision making, a reduction in data development costs as well as the improvement in the quality and predictability of product registrations. There is also consistent interpretation of exposure data by regulatory authorities, a reduced regulatory burden (for both registrants and regulators), and the availability of high-quality data to support health protective regulatory assessments.

References

1. Severn, D. J. Exposure Assessment for Agricultural Chemicals. In *Genetic Toxicology: An Agricultural Perspective*; Fleck, R. A., Hallander, S., Eds.; Plenum Press: New York, 1982; pp 235–242.
2. Hackathorn, D. R.; Eberhart, D. C. Data-Base Proposal for Use in Predicting Mixer-Loader-Applicator Exposure. In *Dermal Exposure Related to Pesticide Use*; Honeycutt, R. C., Zweig, G., Ragsdale, N. N., Eds; ACS Symposium Series, Vol. 273; American Chemical Society: Washington, DC, 1985; pp 341–355.

Chapter 3

Spray Drift Task Force (SDTF)—Development of Data and Models for Determining Off-Target Drift of Agricultural Pesticides

David R. Johnson*

Johnson Management & Consulting, LLC, P.O. Box 509, Macon, Missouri 63552, United States

*Email: davejohn@johnsonmgt.com

In the 1980s, Environmental Protection Agency (EPA) was concerned about predicting and reducing off-target drift from pesticide applications, so they required registrants to submit product-specific spray drift data. However, it was recognized that drift was predominately a function of application parameters such as droplet size, application height, and wind speed rather than the actual product being sprayed. Therefore, the Spray Drift Task Force (SDTF) was formed in 1990 to generate generic drift data applicable to all pesticide products.

SDTF was organized into an Administrative Committee, a Technical Committee, and numerous technical subcommittees that were responsible for directing the development of databases and drift models.

SDTF developed methods and conducted studies on the effects of tank-mix physical properties on droplet size formation, droplet size spectra formed by a wide range of nozzle characteristics, and field drift studies under differing release heights, wind speeds, droplet sizes, and crop canopies. Data were generated for applications by aircraft, ground boom sprayers, airblast sprayers, and chemigation systems. Contract labs and consultants performed the technical activities under the direct supervision of the SDTF, allowing input from the best scientists in the industry and regulatory agencies worldwide working together on a common project.

SDTF was the first major generic task force of this kind and served as a template for subsequent exposure task forces. The SDTF showed that the generic task force approach is an efficient and cost-effective way to generate high-quality non-product-specific (i.e., generic) data for use by both the industry and regulatory agencies.

Introduction

The Spray Drift Task Force (SDTF) was the first major generic task force designed to develop comprehensive technical databases that would be applicable to all agricultural chemical product registrations. The SDTF created procedures for organizing a means by which multiple agricultural chemical registrants could cooperatively pool their technical, business, regulatory, legal, and financial resources into a functional task force that would generate the spray drift data required by EPA. The procedures established by the SDTF served as a template for subsequent generic task forces.

This chapter explains how the SDTF was organized and how it carried out its assignment of generating technical databases and computer models that accurately predict the magnitude of drift from pesticide applications by aircraft, ground boom sprayers, airblast sprayers, and chemigation systems. The resulting data are used to quantitatively determine drift from many types of applications and meteorological conditions, and to show how application equipment can be configured to effectively apply pesticides while minimizing off-target drift.

Regulatory Needs

In the 1980s, EPA was concerned about exposure to water, bystanders, and plants due to off-target drift during the application of pesticides. EPA wanted a mechanism for predicting off-target drift and minimizing the incidence and impact of drift through the use of proper equipment, equipment setup, and application techniques. Consequently, EPA required atomization and drift data for all registered pesticide formulations and end-use products. This was going to result in a multitude of individual product-specific data sets that would outstrip available testing resources.

The problem was that drift is primarily a function of meteorology (wind speed, temperature, and humidity), droplet size, and height of application with very little impact from the characteristics of the formulated product itself. Droplet size was known to be a function of physical chemistry of the spray mix and nozzle parameters (type, size, pressure). This led to discussions between EPA and the pesticide industry on how this dilemma could be addressed.

Solution and Formation of SDTF

The solution was the formation of the SDTF under a Joint Data Development Agreement in April of 1990. It started with 15 companies, that eventually grew to a total of 38 members, plus numerous licensees that contributed to the costs through data compensation settlements.

Data were generated to support the reregistration of more than 2,000 existing products and the registration of future products from SDTF member companies. The purpose of the SDTF studies was to quantify primary spray drift (i.e., the off-site movement of spray droplets at the time of application and before being deposited on the ground outside the target area) from aerial, ground hydraulic, airblast, and chemigation applications. Using a common experimental design, more than 300 applications were made in 10 field studies.

The data generated in the field studies were used to establish quantitative databases which are being used to conduct environmental risk assessments. These databases were also used to validate computer models that are used in lieu of directly accessing the databases.

Besides providing quantitative data on drift, the SDTF studies confirmed that droplet size is the most important factor affecting drift; that the active ingredient does not affect spray drift; that the physical properties of the spray mixture generally have a small effect relative to the combined effects of equipment parameters, application technique, crop canopy and the weather; that spray drift is primarily a generic phenomenon; and that the use of a common set of databases and models for all products is justified.

Prior to initiating the studies, the SDTF consulted with technical experts from research institutions around the world and compiled a list of 2,500 drift-related studies from the scientific literature. Because of differing techniques, it was difficult to compare results across the studies. However, the information from these references was useful in developing test protocols that were consistently followed throughout the testing program

Most of the research was completed by 1995, and the data were reviewed by the Scientific Advisory Panel (SAP) (*1*) in December of 1997. As provided by the Federal Insecticide, Fungicide, and Rodenticide Act (FIFRA), the data submitted to EPA were proprietary to SDTF members and thus allowed the SDTF to seek data compensation payments from non-member companies that relied on the drift data for their registrations. The SDTF remained in existence until 2015 with most of the last 15 years devoted to data compensation negotiations.

Relationship with Regulatory Agencies

Consideration was initially given to allowing the EPA to serve as a member of the SDTF, but this was rejected since EPA's role is to regulate the submission and use of data, and not to be a part of the SDTF itself.

However, a close working relationship between EPA and the SDTF was maintained throughout the planning and implementation of the testing program. The U. S. Department of Agriculture, Environment Canada, and the California Department of Pesticide Regulation were also involved with various aspects of the program.

SDTF Organization

SDTF was organized into three parts:

1. Administrative Committee, with one official representative from each member company that was responsible for managing the business, regulatory, legal, and financial aspects of the Task Force program.
2. Technical Committee, with one official representative from each member company plus other non-official representatives, that was responsible for managing all technical activities.
3. Numerous technical subcommittees, consisting of member representatives with input from outside consultants and contract research groups worldwide with expertise in various technical disciplines such as:

    a. Field studies to design and oversee the actual conduct of field drift studies
    b. Agricultural engineering to direct what application equipment to use in the field drift studies
    c. Analytical chemistry for developing and validating analytical methods and ensuring quality analysis of the samples collected in the field drift studies
    d. Atomization to direct the studies used to measure droplet size spectra
    e. Physical chemistry for analyzing various tank-mix formulations for physical properties
    f. Quality assurance to ensure that all studies were conducted as per the Good Laboratory Practice Standard regulations

g. Report writing for preparing the submissions to the EPA
h. Modelers who developed and validated computer models that would eventually be used for predicting drift under many application and weather conditions
i. Outreach for the production and dissemination of educational materials and publication of peer reviewed papers in scientific journals and meetings
j. Task Force management for coordinating all of the technical activities into a well-organized and functional database by hiring one individual to serve as Study Director of all studies

Conduct of Studies

The technical committees defined the types of studies that were needed for developing the databases. In some cases, the protocols, techniques, and/or research facilities did not exist when the SDTF was formed and had to be developed and validated to the SDTF specifications. This required the utilization of consultants and laboratories located in the United States, Canada, Australia, and England. The actual work was performed by consultants and contract laboratories, which was the only practical way of not overburdening any of the member companies to do the work and to ensure equal costs to all members.

Principles of Drift

Spray tank mixtures are atomized by nozzles into droplets during application. The drop size distribution varies considerably depending upon the nozzle type, nozzle size, nozzle configuration, and speed of the application equipment.

The general principles of drift are:

1. Small droplets have the potential to drift further than large droplets.
2. Droplets discharged at high heights have the potential to drift further than droplets discharged at low heights.
3. Droplets have the potential to drift further under high wind speeds versus low wind speeds.
4. Smaller droplets have the potential to be produced after being discharged from nozzles due to evaporation of water from the droplet under high temperature and low humidity conditions.
5. Droplets only travel in a downwind direction.

Nozzle Effects on Droplet Production

Spray material is broken into a range of droplet sizes as it is discharged from nozzles. This range is referred to as the droplet size spectrum. The whole drop size distribution is important and dependent upon many factors such as:

1. Nozzle type – Different nozzle geometries provide different spray patterns, For example, flood nozzles have a large flow rate that is used when a high volume of spray material is needed. Conversely, disc-core nozzles are used when it is important to produce very small droplets (e.g., ultra-low-volume applications). Air induction nozzles are designed to reduce the volume of very small droplets.

2. Nozzle orifice size - There are multiple sizes within each nozzle type for even greater control of droplet sizes and spray application rates.
3. Nozzle orientation – Nozzles that are oriented straight down will experience more shear and turbulence than nozzles oriented backward, and will thus produce smaller droplets. This is most important at high speeds, such as aircraft applications, as opposed to slower speeds from ground boom sprayers.
4. Air speed – Droplets discharged from a nozzle will experience more shear and produce smaller droplets when discharged from a high-speed turbo-powered fixed-wing plane than from a slower moving helicopter or ground boom sprayer.

Physical Property Effects on Droplet Spectra

The SDTF had to develop techniques and Standard Operating Procedures for determining physical properties (i.e., dynamic surface tension, extensional viscosity, and shear viscosity) of tank mixes. In a large statistically designed study, placebo solutions were produced to represent the total range of physical properties encountered with pesticide tank mixtures (i.e., the actual spray mixtures released through the nozzles as opposed to the physical properties of the formulated product alone).

The full range of possible tank mixtures were tested in wind tunnel facilities for determining the effects of physical properties on droplet formation under a range of nozzle configurations and air speeds at the discharge point of the nozzles. This allowed the SDTF to determine the relationships between physical properties of the tank mixtures, droplet formation, and ultimately drift under a range of environmental conditions.

Dynamic surface tension was the physical property that had the greatest influence on droplet formation. Physical property effects accounted for less than 10% of variability in droplet sizes whereas nozzle parameters accounted for more than 90% of the variability. Formulation type was only significant for Emulsifiable Concentrate and Soluble Liquid formulations applied by high-speed aircrafts. A summary of the relationship between physical properties and atomization are presented in study I95-003 (2).

Effects of Droplet Evaporation on Drift Potential

As droplets are discharged from nozzles, the water part of the droplet begins to evaporate, producing even smaller droplets. Evaporation is a function of temperature, humidity, physical properties, and relative air flow velocities.

A droplet evaporation database was developed from a videography-type study where the rate of evaporation of droplets hanging from small threads under a range of temperatures, humidities, and droplet physical properties was measured. A summary of the methodology and results are presented in Study P92-002 (3).

Measuring Droplet Size Distributions

It is very difficult, and in most cases impossible, to determine the droplet size spectrum as droplets are discharged from sprayers. Therefore, droplet size spectra applicable to aerial and ground boom applications were determined in wind tunnel facilities. Nozzles were placed inside the tunnel where air was forced through the tunnel at speeds that duplicated the air speeds at the nozzle created by aircraft and ground boom sprayers. Laser particle equipment was used to determine the droplet size spectra. All the applications and tank mixes tested in the field were duplicated in wind tunnel studies.

Airblast sprayers commonly contain multiple nozzles of different sizes as a way of further producing a range of droplet sizes for reaching different levels of the crop. This makes it nearly impossible to determine the droplet size spectra that would be discharged from a single sprayer. Therefore, a technique was developed for measuring the total droplet spectra from a sprayer operating outdoors.

It was also impossible to duplicate nozzles used for chemigation applications in a wind tunnel. In this case, a special large facility had to be configured for operating chemigation nozzles indoors where the droplet spectra could be measured.

The droplet size spectra of all equipment and conditions encountered in the field drift studies (i.e., nozzle configurations, air speeds, and physical properties of the materials sprayed) were duplicated in the atomization studies. This approach allowed the SDTF to determine the quantitative relationship between physical properties, nozzle configurations, droplet sizes, air speeds, and drift.

Conduct of Field Drift Studies

All of the field studies consisted of making applications as specified in the protocols. Thick blotting paper-type collectors were placed on the ground at various distances downwind from the application areas to determine ground deposition. Personal air samplers were also placed at various distances for determining the pesticide concentrations in the air. Weather stations were established nearby for determining the meteorological conditions at four different heights. Field studies were conducted to determine drift for the following four types of application.

Aerial Application

Most of the emphasis was on aerial applications because it is the most complicated and has the greatest potential for off-target drift due to the greater height of application, turbulence of aircraft wakes, and usually smaller droplet sizes. A comprehensive survey of aerial applicators was conducted to ensure that the application parameters in the drift studies reflected actual parameters used commercially.

To factor out the meteorological differences among the many application treatments, the SDTF used a covariate experimental design, which is a commonly accepted statistical technique for this type of study. The design entailed a control treatment that was always applied immediately after an experimental treatment. The control treatment was always applied in exactly the same manner, whereas the experimental treatment differed with each application. By using this experimental design, differences between replications of the control treatments were due only to atmospheric conditions since the application procedures were always the same. Differences between experimental treatments were due to both changes in the atmospheric conditions and application procedures. Consequently, differences between experimental and control treatments were due only to application procedures. This allowed direct comparisons to be made among all the experimental treatments by factoring out the effects of weather as measured by the control applications.

This dual application technique was accomplished by a specially-designed plane with two tanks and the ability to hold two different spray booms. The control tank contained diazinon as the analyte and the experimental tank contained malathion as the analyte. This arrangement allowed the plane to apply the experimental treatment immediately after the control treatment without landing the plane. Analytical methods were developed for analyzing both analytes on the deposition cards simultaneously.

Drift was determined from 90 experimental applications and 90 corresponding control applications, which provided an extensive database for evaluating the effect of weather on drift and in validating drift models. A summary of the methodology and results are presented in SDTF Study I94-002 (*4*).

Ground Boom Application

This is the most common type of application, but generally produces less drift than aerial applications because of the lower application heights, slower ground speeds, and usually larger droplet sizes. The slower speeds also eliminated the importance of nozzle orientation. The primary parameters affecting drift are droplet sizes and wind speed.

As with the aerial applications, a statistical covariate design was followed with a constant control treatment applied at the same time as each experimental treatment by using two different ground boom sprayers. A summary of the methodology and results are presented in SDTF Study I94-001 (*5*).

Chemigation Application

This consists of injecting a pesticide into a center pivot irrigation system. Thus the pesticide was applied as part of the water discharged from the irrigation nozzles. The SDTF measured drift from the following two types of systems:

1. A high-pressure system operated at 70 pounds per square inch with impact-type nozzles located on top of the pivot 12 feet above the ground
2. A low-pressure system operated at 20 pounds per square inch with drop rotary spinner-type nozzles located 5 feet above the ground

Both systems were tested with and without an end-gun. A summary of the methodology and results are presented in SDTF Study F94-022 (*6*).

Airblast Application

Airblast sprayers are used for applying pesticides to orchard and trellis crops. Applications are made in and around plant canopies rather than in open areas as with the other three types of application. With airblast applications, the characteristics of the crop canopy is the most important parameter affecting drift. Therefore, field drift studies were conducted on the following four types of crop canopies.

1. Foliated and dormant deciduous fruit trees represent the most common type of orchard crop and have relatively open canopies that allow some of the droplets to penetrate through the foliage.
2. Pecan and almond trees have an open canopy and are much taller than the deciduous trees.
3. Orange and grapefruit trees have dense canopies that reduce droplet movement through the foliage, resulting in more than 99% of the spray material being intercepted by the foliage. Data were collected for applications by low-volume/small droplet mist blowers as well as standard airblast sprayers.

4. Grape vineyards are much shorter than the orchard crops. Drift was determined for both airblast sprayers and wrap-around (directed) sprayers used on trellis crops.

In addition to measuring downwind ground deposition, vertical strings were positioned between rows and outside the deciduous, citrus, and grape crops for measuring the vertical deposition profiles. This provided more detailed information about the movement of droplets within and outside the canopies. Due to the canopy effects, drift from orchards and vineyards was less than from aerial and ground boom applications. A summary of the methodology and results are presented in SDTF Study I95-004 (7).

Modeling

The SDTF data verified conventional knowledge that spray drift is a predictable phenomenon based on droplet size spectra, application height, and meteorological conditions. Due to the complexity of evaluating all possible interactions of the numerous application variables, a computer model is the most practical way to conduct environmental risk assessments that would allow a meaningful quantitative estimate of drift under virtually all types of conditions.

For aerial applications, a highly sophisticated simulation model had been developed previously by the U. S. Department of Agriculture (USDA) Forest Service for forestry applications. The SDTF, EPA and USDA worked together under a Cooperative Research and Development Agreement (CRADA) to adapt and validate this model for agricultural applications using the data generated in the SDTF field and atomization studies.

The SDTF version of the subsequent model included ground and orchard empirical models and was named AgDRIFT. The bulk of the research was completed in 1995, although the models continued to be refined in subsequent years and were posted on EPA's website.

Outreach Activities

The primary SDTF goal was to produce databases and models for use by EPA and other regulatory agencies for determining off-target drift potential. The resulting data also had value to educational and other scientific programs. Therefore, the SDTF undertook the following outreach activities.

1. Scientists who worked on the SDTF project continued to improve the databases and models, published several papers in refereed journals, and made numerous presentations to various stakeholder meetings.
2. The SDTF produced tens of thousands of educational booklets that presented the results in simple language. Booklets were produced for each of the four types of application plus a booklet on atomization. With the common use of the internet, these booklets were placed online for easy access by the public. Copies of the booklets are available from the author upon request.
3. Generic protocols were released for use by other research groups.
4. Working with American Society of Agricultural and Biological Engineering (ASABE), a national standard was put in place to simplify the description of spray drop size classes for use with the drift model and for use in pesticide label improvements.

Organization of Generic Task Forces

The SDTF approach was duplicated successfully by subsequent generic task forces by:

1. Pulling together the best scientists in industry and the regulatory agencies to work cooperatively on all aspects of a project
2. Organizing the task force into administrative and technical committees, along with various subcommittees to manage the technical aspects of the task force
3. Using technical subcommittees to manage studies that were actually conducted by professional laboratories and consultants rather than relying on the individual member companies
4. Developing new testing procedures when current technology was not sufficient for accomplishing the goals and objectives of the research program
5. Establishing a close relationship between the task force and regulatory agencies to ensure the planning and execution of studies was acceptable to the registrants and regulators for making regulatory decisions
6. Making results available to other research and educational programs
7. Resulting in superior databases with wide applicability that were developed in a very effective and cost-efficient manner

Summary and Conclusion

The SDTF developed spray drift models and databases that are being used by the agricultural chemical industry and the EPA in environmental risk assessments. Even though active ingredients do not differ in drift potential, they can differ in the potential to cause adverse environmental effects. Since drift cannot be completely eliminated with current technology, the SDTF databases and models are being used to determine if the drift from each agricultural product is low enough to avoid harmful environmental effects. When drift cannot be reduced to low enough levels through altering equipment setup and application techniques, buffer zones are often imposed to protect sensitive areas downwind of applications.

Being the first major generic task force designed to produce a database that could be used by the industry and regulatory agencies or virtually all types of products and application methods, the SDTF led the way in producing a mechanism that would serve as a template for subsequent task forces.

The generic task force approach enables the industry to produce a single, well-organized and comprehensive database that is easy to use and is applicable to virtually all product registrations. It would have been impossible to produce a usable database by combining data from a multitude of individual product-specific studies. It also would have been impossible for any one company alone to produce such a comprehensive database. By combining technical and financial resources, the database was produced at a relatively low cost to each company (approximately $25 million spent by the SDTF versus a 1990 estimate of nearly $1 billion by any other approach). This is an example of a win-win-win project that benefited the industry, regulatory agencies, the scientific community, and other stakeholders at large.

As a side benefit, both the regulatory and industry personnel participating on this project gained invaluable knowledge from the interactions and exposure to so many other scientists and professionals.

Acknowledgments

The author acknowledges the many representatives of the SDTF member companies who directed the work of the Task Force, representatives of EPA who provided invaluable inputs that ensured that the final databases and models would satisfy their regulatory needs, and the consultants and laboratories that generated the data and models. These groups not only generated the data to meet the objectives of the SDTF, but also contributed to the scientific community at large through the generation of new information and development of new techniques.

References

1. U. S. Environmental Protection Agency, Aerial Spray Drift Review by Scientific Advisory Panel Meeting, December 11, 1997.
2. Hermansky, C.; Hewitt, A. Relationship Between Physical Properties and Atomization: Integration and Summary, SDTF Study I95-003, January 27, 1999, EPA MRID 44747401.
3. Hill, R. L. Spray Drift Task Force Droplet Evaporation of Spray Drift Test Substances, SDTF Study P92-002, September 24, 1996, EPA MRID 44134101.
4. Johnson, D. R. Drift from Applications with Aerial Sprayers: Integration and Summary of 1992 and 1993 Field Studies, SDTF Study I94-002, May 17, 1995, EPA MRID 43803501.
5. Johnson, D. R. Drift from Applications with Ground Hydraulic Sprayers: Integration and Summary of 1992 and 1993 Field Studies, SDTF Study I94-001, January 10, 1995, EPA MRID 43508001.
6. Johnson, D. R. Spray Drift Task Force 1994 Field Chemigation Study, SDTF Study F94-022, November 6, 1995, EPA MRID 43845901.
7. Johnson, D. R. Drift from Orchard Airblast Applications: Integration and Summary of 1993 and 1994 Field Studies, SDTF Study I95-004, December 5, 1995, EPA MRID 43925701.

Chapter 4

Agricultural Handler Exposure Task Force—Development of a Generic Exposure Database for Agricultural Workers Performing Mixer/Loader and Applicator Tasks

David E. Barnekow*

AHETF Administrative Committee Chair
Corteva Agriscience LLC, Indianapolis, Indiana 46268, United States

***Email: david.barnekow@corteva.com**

The Agricultural Handler Exposure Task Force (AHETF) was established in 2001 to generate agricultural pesticide handler exposure data meeting U.S. Environmental Protection Agency (EPA) data requirements to support product registrations. EPA had explicitly imposed data requirements on several pesticide registrants through data call-in notices, reregistration requirements, or through consultation with applicants regarding prospective registrations. These companies, and eventually many others, agreed to form AHETF and jointly develop the reliable generic data necessary to support the continued EPA registration of most products approved for agricultural uses.

The primary AHETF goal was to collect worker exposure monitoring data for a wide variety of pesticide mixing and loading scenarios (such as mixing/loading different formulation types and mixing/loading systems), application scenarios using different types of application equipment (such as ground boom, airblast, aerial, and handheld sprayers) and engineering controls (such as enclosed cabs). These pesticide handler exposure data replaced data from Pesticide Handler Exposure Database (PHED, EPA 1998) that had been found to suffer from various scientific deficiencies by EPA and other regulatory agencies. AHETF submitted reports and the Agricultural Handler Exposure Database (AHED®) to EPA, California Department of Pesticide Regulation (CDPR), Canada's Pest Management Regulatory Agency (PMRA) and U.S. Department of Agriculture (USDA) (collectively called the Joint Regulatory Committee, or JRC) that was involved with the overall development of the testing program.

Introduction

The focus of this chapter is on the establishment and structure of the Agricultural Handler Exposure Task Force (AHETF), the technical need for generic data to describe exposure potential for pesticide-handling situations in North American agriculture, several principles of generic data and how AHETF determined the scope of its testing program which includes 25 distinct scenarios. The chapter by Eric Bruce will provide a more detailed description of the technical aspects of exposure data generation, analysis and reporting.

Formation of the Agricultural Handler Exposure Task Force

Based on the recognition that handler exposure to pesticides is generic rather than product-specific, the industry and the regulatory agencies cooperated in establishing the Pesticide Handler Exposure Database (PHED) in the early 1990s. Companies contributed exposure data without any proprietary or compensation protections. This handler exposure database served as the foundation for many exposure and risk assessments for occupational pesticide handlers (also known as operators).

By 2001, several North American pesticide regulatory authorities; U.S. Environmental Protection Agency (EPA), Canada's Pest Management Regulatory Agency (PMRA) and the California Department of Pesticide Regulation (CDPR) identified a need to address the significant technical limitations of PHED (*1*). The technical limitations of PHED included inadequacies in body part measurement and quality assessment of data; lack of entire body dermal estimates, and minimal diversity and representativeness of handler conditions. To address the inadequacies of the existing database EPA announced they would compile all new available exposure data into scenarios to form a new exposure database for supporting operator exposure risk assessments. With limited new studies available EPA began requesting new handler exposure studies from individual companies during new section 3 submissions and registration reviews of existing active ingredients.

In response to broad pressure on the industry by the EPA to develop new and more robust worker exposure data, the need for a generic Task Force was recognized by the industry and supported by the Agency. The industry decided to participate in this process by establishing the AHETF in December 2001 with the mission of sharing resources in the design, evaluation and development of a proprietary agricultural mixer/loader and applicator exposure database for use in regulatory risk assessments. By forming a Joint Data Development and Limited Liability Company (the Task Force), the industry hoped to develop a broader database with more robust data for many scenarios since there were insufficient high quality existing data to allow for adequate operator risk assessment. AHETF data were developed to meet pesticide registration requirements of the Federal Insecticide and Fungicide Act (FIFRA). Therefore, AHETF data are considered proprietary to AHETF member companies and subject to the data compensation provisions of FIFRA.

The primary AHETF goal was to collect worker exposure monitoring data for a wide variety of pesticide handler scenarios, such as mixing/loading several different formulation types and application of pesticides using various types of application equipment and engineering controls. These pesticide handler exposure data replaced data from PHED (2) that had been found to suffer from various scientific deficiencies by EPA and other regulators.

Structure and General Operation of the Agricultural Handler Exposure Task Force

The AHETF consists of manufacturers and formulators of agricultural pesticides. Representatives from the 27 member companies are available to participate in managing the Task Force through two main committees, the Administrative and the Technical. Member companies of the AHETF are:

Aceto Life Sciences, LLC	ADAMA
Amvac Chemical Corporation	Arysta LifeScience North America
Atticus, LLC	BASF Corporation
Bayer CropScience	Dow AgroSciences LLC
Drexel Chemical Company	DuPont Agricultural Products
FMC Corporation	Gharda Chemicals International, Inc.
Gowan Company	Helena Agri-Enterprises, LLC
ISK Biosciences Corporation	K-I Chemical USA
Loveland Products	Luxembourg-Pamol, Inc.
Mitsui Chemical Agro, Inc.	Monsanto Company
Nichino America, Inc.	Nippon Soda Co., Ltd
Nufarm Americas Ltd.	PBI/Gordon Corporation
Syngenta Crop Protection	Taminco US LLC
Valent USA Corporation	

Administrative Committee

The business and affairs of the Task Force were managed by an Administrative Committee that consisted of one representative from each member company. A Chair of the Administrative Committee, elected by the committee members, has overall responsibility for carrying out the actions of the Task Force.

The Administrative Committee was responsible for managing the work of the Task Force, including supervising and managing the acquisition, reference, and development of Task Force data, engaging consultants and attorneys as necessary, managing financial matters, adopting procedures to develop and record information and otherwise to conduct the work of the Task Force in a manner that supported the data compensation rights of the Task Force and its members.

Technical Committee

The Technical Committee was responsible for the general supervision and management of the acquisition or development of generic agricultural handler exposure data and all technical matters. The activities were subject to the control of the Administrative Committee. Each member company appointed one voting representative to the Technical Committee, although companies were encouraged to designate additional non-voting participants. The Technical Committee was led by a Chair of the Technical Committee that was elected by the Technical Committee members.

The Technical Committee executed its responsibilities for the acquisition or development of generic agricultural handler exposure data largely through the formation of subcommittees as deemed necessary and appropriate to accomplish the purposes of the Task Force. Subcommittees served only an advisory function; final decisions on matters considered by any subcommittee were made by the Technical Committee.

Data analysis, analytical and field studies subcommittees were three main subcommittees the Technical Committee relied upon throughout data acquisition and data development. Through the advisement of these subcommittees the Technical Committee chose appropriate test materials, prepared protocols, conducted analyses, received and evaluated protocols from testing laboratories, and reviewed and evaluated progress and final reports from such laboratories. The subcommittees developed the underlying principles and procedures for conducting data development. They developed standard operating procedures (SOPs) for analytical and field work, a Governing Document describing the need for human exposure data and design principles for how AHETF would develop exposure data, specific plans for individual studies, and acceptance criteria by which existing studies were evaluated for inclusion into the new generic agricultural handler exposure database. Subcommittees oversaw the performance of contract research organizations (CROs) which were contracted to perform technical work for the Task Force, and developed any technical positions advanced by the Task Force in response to inquiries by EPA or any other regulating authorities. The recommendations of contractors were reviewed and approved by the Technical Committee, subject to final approval of the Administrative Committee. For Task Force work subject to EPA's Good Laboratory Procedures regulations (GLPs), the Technical Committee selected only those testing laboratories that were in compliance with GLPs. The Technical Committee was also responsible for addressing data storage, data archiving, and all other GLP requirements pertaining to study sponsors. To assist in overseeing GLP compliance a Quality Assurance subcommittee was formed.

Engagement with Regulatory Authorities

From the formation of the AHETF it was recognized that an effective collaboration with the key North American regulatory authorities was essential to the development of a robust generic handler exposure database. The Joint Regulatory Committee (JRC), was setup with AHETF member representatives and representatives from EPA, CDPR, PMRA and the U.S. Department of Agricultural (USDA) as a forum for the agencies to provide input to AHETF on science and policy matters including the planning, development and analysis of the exposure monitoring studies.

Initial efforts of the JRC focused on considering existing handler exposure studies for inclusion into the new generic database. With JRC consultation, inclusion criteria were established for evaluating existing study data acceptability. AHETF consultants developed summary reports for each study evaluating each of the criteria for acceptance. These summary reports, known as Secondary Review Reports (SRRs), were reviewed with the JRC for their recommendation on each study's acceptance. As acceptable monitoring data were identified, AHETF acquired the rights to cite the data from the study owner so it could be included in the new generic database.

Following the acquisition of acceptable existing monitoring data AHETF developed scenario plans that identified where additional monitoring data were needed to complete the scenario. These plans were reviewed and approved by the JRC before being implemented by AHETF.

External Engagement of Agricultural Experts

AHETF recognized early on the need to engage external experts to aid in the identification and development of the scenario plans of the Task Force. These external experts included state agricultural extension specialists, agricultural equipment manufacturers, commodity / use site organizations (e.g., Cotton Growers Association, Vegetation Management Association, Pecan Growers Association) and pesticide applicator associations (i.e., pilot and application company owners and the National Agricultural Aviation Association). These associations provided valuable insight into how specific handling activities were performed, what equipment configurations were commonly used, and typical product formulations used. They also assisted in characterizing equipment or practices variability across geographic regions, product types, use sites, etc.

AHETF formed the Agricultural Handler Advisory Panel (AHAP). The AHAP included member company representatives and representatives from the Florida Fruit & Vegetable Association, National Cotton Council, National Agricultural Aviation Association, American Nursery and Landscape Association, USDA, American Farm Bureau, Michigan State, Northwest Horticultural Association, and CropLife America. This group was instrumental in guiding the prioritization of testing and provided technical information about enclosed cabs, backpacks, greenhouse applications, and seed treatment. The group was also consulted about existing studies to confirm if equipment and practices represented in the existing studies were still commonly in use.

Following the implementation of the Protection of Human Subjects rule (3), a Technical Advisory Panel (TAP) was formed for the purpose of soliciting information from commodity groups, educators, contract researchers, and university extension personnel to develop specific testing criteria to be incorporated into scenario plans. In addition to the TAP, AHETF consulted with an external ethicist for guidance on the resolution of ethical concerns raised by the Human Studies Review Board (HSRB). Concerns focused on setting criteria for the inclusion or exclusion of potential participants and ethical procedures for engagement with participants. Some of the important concerns were around native language for communication, limited literacy and inclusion of women (especially ensuring the prohibition of pregnant and nursing women as participants). To address these concerns and others, AHETF developed new or expanded SOPs to address these criteria. These SOPs were reviewed and approved by the EPA and the HSRB before being utilized in the conduct of a monitoring study initiated after the implementation of the Protection of Human Subjects rule.

AHETF also utilized survey tools to collect information to identify and characterize parameters to be included into the scenario design, such as, application or mixing/loading conditions, crops /application areas and the selection of geographic regions for testing. These surveys will be discussed later in this chapter.

Technical Needs for Generic Data

Regulatory agencies found that certain data supporting pesticide registrations for a particular active ingredient may be applicable to many pesticide products. In these cases, the data and their utility are considered "generic" in nature. This includes worker exposure data for handling pesticides and there is a long history of using generic worker exposure data during initial pesticide exposure assessments. For compounds with low volatility (which includes most agricultural pesticides), an accepted generic principle is:

If all other conditions are similar, the magnitude of exposure does not depend on the particular active ingredient used.

The practical importance of this generic principle is that it permits a monitoring unit (MU) based on one surrogate chemical to be used to predict exposure to other active ingredients under a similar set of handling conditions. However, for a data set to be valid and useful for generic use by registrants and regulatory agencies, some fundamental conditions must be met, including:

- Use of appropriate surrogate active ingredients (e.g., stable with good analytical methods)
- Use of appropriate "end-use" products (e.g., formulation types that are common)

- Tasks, equipment, and handling practices that are representative
- Clothing and personal protective equipment (PPE) conditions that are supported by product labels
- Accurate measurement of a condition(s) or parameter(s) that is proportional to contact with active ingredient (i.e., a normalizing factor such as the amount of active ingredient handled, AaiH)

This section describes how generic data are used by regulatory agencies and some of the principles that AHETF followed to ensure its worker exposure data for a variety of pesticide-handling scenarios would be considered generic and useful for regulatory exposure assessments.

Historical Use of Generic Data for Regulatory Exposure Assessments

Exposure is not determined by the properties of the active ingredient in a formulated product but rather by physical factors and activities involved in handling the product. Such approaches have been accepted since the 1980s (4). Using this principle, a database of exposures generated from a specific set of surrogate active ingredients can be used to predict exposure occurring from any active ingredient in a product that is used under conditions similar to the surrogates. Thus, generic exposure databases organize exposure measurements in a manner that represent distinct exposure situations, or "scenarios". Regulatory agencies and others can use generic exposure data collected for a particular scenario to predict exposures to all handlers performing the work activities described by that scenario.

The concept of using a generic database has been discussed and supported at several Scientific Advisory Panel (SAP) and HSRB Meetings (5–9). This broadly accepted, generic approach for conducting worker exposure assessments is supported by regulatory authorities throughout North America (e.g., EPA and Health Canada) and Europe (e.g., Organization for Economic Cooperation and Development) (5–7). Support for this approach was reaffirmed by the January 2007 SAP for use in agricultural handler exposure assessments (3).

Principles of Generic Data

When estimating exposure to workers who handle pesticides, a major factor to address is that several parameters can contribute to the likelihood and level of exposure. These factors include: handling liquids versus solids, product packaging type, using open versus closed systems, applying products with various equipment types, using personal protective equipment (PPE), and whether the worker mixes/loads or applies or does both. The number of parameters makes it impractical to generate human exposure data for all potential situations, so a number of simplifying steps have been adopted. Most importantly, a discrete number of these combinations are defined and a set of data is generated to represent each of these handler scenarios. A pesticide-handling scenario is a well-defined worker exposure situation, usually characterized by specific agricultural chemical handling tasks and equipment. When conducting exposure assessments for a pesticide product, registrants and regulatory authorities define the various scenarios that would be involved based on factors such as the formulation type, expected application techniques, and label requirements (or restrictions) for engineering controls, clothing/PPE, crop uses, use rates, etc. In particular, it was desired that AHETF data would generally characterize exposures resulting from the future use of a pesticide active ingredient assuming some quantifiable amount of active ingredient contact (e.g., AaiH). These characterizations of exposure will be useful to inform future reviews of pesticide active ingredient uses since for a given scenario exposure potential is generally proportional to the amount of chemical handled.

For the purposes of the AHETF monitoring program, the basic element of a scenario was considered the handler-day (HD). Each handler-day corresponded to a particular worker and the scenario-related activities performed during a single workday. Implicitly associated with each HD was a complex set of conditions. The number of components included, but was not limited to worker behavior, active ingredient used, amount of chemical handled, tasks performed, and numerous environmental factors.

A monitoring unit (MU) was the experimental tool used by the AHETF monitoring program to predict exposures. An MU was a set of scenario-specific handler-day conditions that were experimentally 'selected' (chosen, scripted, or otherwise constructed) to represent expected future HD conditions. Each MU was experimentally monitored to yield a set of exposure measurements. Thus, each MU provided a measurement of the actual exposure resulting from the selected HD conditions. The MU provided a predicted future exposure if the handling conditions used for the MU are similar to future HD conditions.

In addition to generating a set of measurements for discrete scenarios, another important principle of generic worker exposure data was that the set of measurements (or MUs) would be most useful in a generic sense if a variety of conditions that might impact exposure levels were appropriately chosen to represent current conditions and diversified to include common differences occurring within the scenario. This included the items mentioned previously: active ingredients, products, tasks, equipment, handling practices, clothing, PPE, and some measure of contact potential such as AaiH.

Defining the Scope of the AHETF Testing Program

AHETF defined a scope that included 25 pesticide-handling situations. Nearly all of these were supported by exposure data that met strict acceptability requirements and were generated by AHETF or acquired for citation by AHETF members. A few were represented by data for another scenario that was determined to have similar, or lower, exposure potential.

Member and Non-member Surveys

One of the first AHETF efforts was to determine the scope of its monitoring program which involved listing the specific pesticide-handling scenarios that would need to be covered by an appropriate set of monitoring data. The existing list of scenarios in PHED was taken as a starting point, but it was recognized that some scenarios might best be redefined while others might already be represented by adequate data or might not be applicable to an agricultural setting (e.g., spraying paint). AHETF and JRC agreed that mixing/loading situations, when practical, should be monitored separately from application situations since there were too many combinations to do mixer/loader/applicator (MLA) studies for all combinations and because these tasks are sometimes performed by separate workers. However, it was recognized that some situations were so commonly conducted by the same person that MLA might be important, such as backpack applications of liquids.

In January 2002, AHETF distributed a survey to registrants, including non-member companies. The survey provided a list of major application scenarios relevant to the agricultural industry. The registrants were asked to indicate the importance of these categories to their business and to list other scenarios they felt were important. During this process, several application types were recommended to be out of scope because the majority of respondents indicated they were not important to their business or were not applicable to an agricultural setting (e.g., applications to livestock). Based on the survey results and subsequent discussions with the JRC, AHETF developed a list of pesticide-handling scenarios under consideration for its scope. For important application types, AHETF considered when/if various engineering controls should be included as separate scenarios (e.g. open cabs versus closed cabs). AHETF evaluated the common formulation types and developed a list of mixing/loading scenarios to be included. Also, AHETF considered when combined MLA studies would be most appropriate. Including a scenario within the scope of AHETF did not necessarily mean AHETF would generate new data for the scenario since it was AHETF's policy to consider purchasing/citing existing data and/or using data from one scenario as a surrogate for another scenario. The policy of considering all available exposure data meeting the requirements for inclusion in the generic database before intentional exposing additional workers was critical to complying with the ethical standards set under the Human Studies Rule (3).

Initial Scope of 33 Scenarios

Based on the survey results and subsequent discussions with the JRC, AHETF developed a list of pesticide-handling scenarios in 2005 that were still under consideration for its scope. In addition, AHETF recognized that some scenarios could potentially be combined during the testing program.

Prioritization of Scenario Testing

After developing the proposed list of 33 scenarios, AHETF surveyed its current members to prioritize the scenarios to indicate which scenarios were most important to their company. The purpose was to provide guidance on testing priorities.

Final List of Scenarios Addressed by AHETF

By 2008, AHETF had completed the review of (non-seed treatment) existing studies and understood what suitable data were available and which scenarios would likely need additional data. In addition, the post-rule environment was better defined in terms of how to plan a scenario and how to determine the number of MUs that would likely be needed to adequately represent a scenario. Subsequently, AHETF reviewed existing seed treatment studies and determined that many were suitable and available for a generic data set and additional AHETF data were not needed.

As post-rule scenario plans were developed and seed treatment studies were reviewed, refinements were made to the scope which resulted in a final list of 25 scenarios.

The following 15 scenarios were unchanged from the original list of 33:

Mixer/Loader Scenarios:

1. Open pour mixing/loading of liquids (OPML)
2. Open pour mixing/loading of dry flowables (DFML)
3. Open pour mixing/loading of wettable powders (WP)
4. Open pour mixing/loading of water soluble packets (WSP)
5. Open pour loading of granules (OPG)
6. Closed system loading of liquids (CSLL, later renamed MTL)
7. Closed system loading of granules (i.e., solid formulations, CLS)
8. Chemigation mixing/loading (ChemML)

Applicator Scenarios:

9. Open cab airblast application of liquids (OCAB)
10. Closed cab airblast application of liquids (CCAB)

Mixer/Loader/Applicator Scenarios:

11. Backpack mixing/loading/application of liquids (BP-L)
12. Backpack loading/application of granules (BP-G)

Seed Treatment:

13. On-farm seed treatment with solids (OFST-S)
14. On-farm seed treatment with liquids (OFST-L)
15. Loading and planting treated seeds (LPTS)

The following 6 scenarios resulted from splitting a scenario from the original list of 33 into more than one distinct scenario:

Commercial Seed Treatment Was Split into:

16. Commercial seed treatment, cleanout (CST-C)
17. Commercial seed treatment, packaging (CST-P)
18. Commercial seed treatment, treating/multi-activity (CST-TM)
19. Commercial seed treatment of small-seeded vegetables (CST-SSV)

Rights-of-Way Application of Liquids Was Split into:

20. Backpack application of liquid sprays in utilities rights-of-way (BP-ROW)
21. Handgun application of liquid sprays in utilities rights-of-way (HG-ROW)

The Following Four Scenarios Are Combinations of Original Scenarios:

22. Closed cockpit aerial application of liquids (CCA): combined fixed-wing and rotary-wing application of liquids
23. Open cab ground application of liquids (OCG-L): combined with soil incorporation and without soil incorporation
24. Closed cab ground application of liquids (CCG-L): combined enclosed cab ground boom and enclosed cab airblast application of liquids (a separate closed cab airblast monograph was also submitted, but no separate closed cab ground boom monograph was ever submitted)
25. Mixing, loading and application of pesticides in managed horticultural facilities using powered handgun equipment (Greenhouse/Nurseries, GH/N): combined low pressure and high pressure handwand applications of liquids

The following 9 scenarios from the original list of 33 were not addressed by AHETF:
Applicator scenarios:

1. Open cockpit rotary application of liquids
2. Enclosed cockpit fixed-wing application of granules

3. Enclosed cockpit rotary application of granules
4. Open cockpit rotary application of granules
5. Enclosed cab ground boom application of granules
6. Open cab ground boom application of granules
7. Open cab ground boom application of granules with soil incorporation

Mixer/loader/applicator (MLA) scenarios:

8. Belly grinder MLA of granules
9. Handheld Mist blower / Fogger MLA of liquids

Other Supporting Surveys and Reports

In addition to the handler exposure reports and monographs that were the primary focus of AHETF, several additional projects were undertaken that provided support for exposure or risk assessments that utilize AHETF exposure results. These include:

- Chemigation Expert Survey (CES) Report: summarizes two surveys of chemigation experts (one for on-farm and one for greenhouse/nursery situations) in various areas of the U.S. and concludes that important factors in worker exposure such as formulation type, equipment used, mixing/loading procedures, work cycles, total work time, etc., used in chemigation are essentially similar to those for other mixer/loader scenarios that have exposure data in the AHETF generic database. Therefore, no additional exposure estimates are required to address chemigation mixing/loading exposure potential. Worker exposure data are not needed for application by chemigation since workers are not allowed in the treatment area while irrigation is being conducted.
- Chemigation Acres/Day Using Drip Irrigation Survey: summarizes a survey of 158 growers of various crops that quantifies the number of acres per day that are typically chemigated using drip irrigation systems on U.S. farms.
- Characterization of the Seed Treatment Industry: summarizes a survey designed to characterize the seed treatment industry in the U.S. and Canada. The resulting survey report provides information on seed treatment systems, seed treatment capacity, packaging, and worker tasks for a variety of seed types that are commonly treated with pesticides.
- Commercial Seed Treatment Survey: summarizes a survey of commercial seed treatment facilities in the United States and Canada to collect information about the amount of seed treated per day for various seed types. The survey report proposes standard values to be used in exposure and risk assessments for various seed treatment activities, including daily amount of seed treated (by seed type and facility type), duration of the seed treatment season (i.e., days of treatment per year) for various seed and facility types, and frequency and duration of equipment cleanout activities for various seed and facility types.
- Small-Seeded Vegetables (SSV) Survey: summarizes information from multiple seed treatment facilities regarding the typical treatment processes, production capacity, and amount of seed typically handled per person per day for a wide range of SSVs. A survey report provides information to support exposure and risk assessments for SSV uses, most likely based on AHETF exposure data for the commercial seed treatment scenarios.

- Greenhouse and Nursery Application Volume: summarizes interviews of 613 greenhouse and nursery growers who use either a backpack or handgun sprayer to characterize several parameters related to worker exposure and risk assessments. The report supports lowering the default assumptions for number of gallons one worker treats per day while making pesticide applications with a backpack or handgun.
- Survey of Acres Treated per Day by Airblast Sprayers: collected acres/day information from growers and commercial applicators seven states representing 20 major airblast-treated crops. The survey report summarizes acres/day information that could potentially be used in deterministic or probabilistic exposure assessments for occupational airblast applicator exposure.

AHETF also conducted a variety of scenario-specific surveys that supported the design of various handler scenarios. They include:

- Survey of Handheld Equipment Usage in Greenhouses and Nurseries (supported the greenhouse and nursery handheld applications study)
- Survey of Handheld Equipment Usage in U.S. and Canadian Vegetable Greenhouses (supported the greenhouse and nursery handheld applications study)
- Survey of the Use of Open Cab Tractors for Soil Incorporation of Pesticides (identified high use areas associated with this application scenario)
- Application Insights in Integrated Vegetation Management in ROW Applications (supported the right-of-way applications study)
- Survey of Backpack and Handgun Usage in Utility Rights-of-Way (supported the right-of-way application study)
- Survey of Closed System Liquid Loading Systems for Small Containers (supported the closed loading of liquids study)

Exposure Monitoring Data Reporting, Review and Regulatory Use

At the conclusion of the field and analytical data collection reports were prepared for submission to three Regulatory Authorities of the JRC. There were some changes over time regarding what a study involved in terms of MUs and handling scenarios; accordingly, contents of study reports changed slightly. Most importantly, studies changed from covering a single monitoring area to including all areas needed for an entire scenario. These changes took place when the Human Studies Rule was implemented and when AHETF began producing monographs for completed scenarios.

For studies conducted Pre-Human Studies Rule, a single summary report was written. Each summary report included; a summary sections for the entire study, field phase, and analytical phase, a summary sections for all numeric results, summary tables for individual replicate (now referred to as a MU) information and results (included field fortification data, adjusted residues, unadjusted residues, and total dermal exposure [TDE] calculations). Appended to this summary report was the Field Report and Analytical Report. These initial summary reports were designed to provide final results for every replicate of exposure monitoring following standard calculation methods that AHETF planned to use for all replicates (whether purchased studies or AHETF-conducted studies).

Following the implementation of the Human Studies Rule, AHETF changes the way reports were assembled. Adjusted residues were no longer reported in the study report (this would be done in monographs). The study report was expanded to include all the details of the Field Report, so an appendix for a field report was discontinued and an extensive new section was added to discuss the process used for recruiting employers and subjects.

AHETF generated scenario monographs for the benefit of regulators and other potential database users. Each monograph includes a description of the scenario, a summary of all the MUs representing the scenario, and an assessment of the data adequacy for that scenario. The monograph detailed the calculation of the normalized total dermal exposure (nTDE) and normalized total inhalation exposure (nTIE) for every MU and the application of adjustments (field fortifications, Face PPE worn, and face/neck exposure to whole head exposure).

One specific requirement of the Human Studies Rule was to report all correspondence with the independent review board (IRB) for each study reviewed/approved by the EPA and the HSRB. For those studies, a new report was prepared and submitted – these reports were called "IRB Correspondence Reports". They presented all correspondence in chronological order, plus tables that summarize the dates of submission and approval for all stages of the study (e.g., original submission, amendments, deviations, annual reviews, study close-out reports, etc.). For studies conducted in California, these reports also included correspondence with CDPR.

Following completion, each study was submitted to all three regulatory authorities, EPA, PMRA and CDPR for their review and acceptance. In the case of the non-observational post-rule studies submitted to the EPA, before EPA could utilize the study an HSRB review was required.

EPA will prepare a summary presentation of the study that describes the need for the data, the appropriateness of the scientific methods utilized, and how compliance with the ethics requirements were satisfied. EPA asked the HSRB to respond to two basic charge questions:

1. Was the research scientifically sound and likely to provide reliable data useful for assessment?
2. Did the available information support a determination that the study was conducted in substantial compliance with the applicable requirements of 40 CFR part 26?

Following concurrence from the HSRB on the two charge question the study data were acceptable for EPA to use in handler risk assessment.

Today the generic exposure database for agricultural workers performing mixer/loader and applicator tasks developed by the AHETF is utilized by EPA, PMRA and CDPR for their handler risk assessment.

Acknowledgments

Thank you to the 27 AHETF member companies and their Administrative representatives for providing leadership and the technical representatives for the countless hours invested over the past 20 years to design and develop a robust agricultural pesticide handler exposure database, to the JRC members whose guidance was indispensable, to the contract laboratories and consultants who did the actual hands-on work, and finally to the Task Force Manager, Dr. Dave Johnson; Task Force Administrative Assistant, Rhonda Bichsel; and the Task Force study directors, Dr. Doug Baugher and Eric Bruce.

References

1. Stasikowski, M. *Revisions to PHED, Memorandum to Daniel P. Fay of the EPA*, Washington, DC, March 16, 2001.

2. *EPA, PHED SURROGATE EXPOSURE GUIDE - Estimates of Worker Exposure from The Pesticide Handler Exposure Database Version 1.1*, August 1998.
3. *Federal Register*: February 6, 2006, Volume 71, Number 24, Pages 6137–6176, 2006, govinfo.gov/content/pkg/FR-2006-02-06/pdf/FR-2006-02-06.pdf (accessed March 29, 2021).
4. Hackathorn, D. R.; Eberhart, D. C. Data Base Proposal for Use in Predicting Mixer-Loader-Applicator Exposure. In *Dermal Exposure Related to Pesticide Use*; Honeycutt, R., Zweig, G., Ragsdale, N., Eds.; ACS Symposium Series 273; American Chemical Society: Washington, DC, 1985, pp 341–355.
5. *EPA, Health Canada, California Environmental Protection Agency – Review of Worker Exposure Assessment Methods*, FIFRA Scientific Advisory Panel, Washington DC, January 9–12, 2007.
6. *EPA, A Set of Scientific Issues Being Considered by the Environmental Protection Agency Regarding: Review of Worker Exposure Assessment Methods*, FIFRA Scientific Advisory Panel Meeting, SAP Minutes No. 2007-03, Washington DC, January 9–12, 2007.
7. *EPA, Health Canada, California Environmental Protection Agency – Scientific Issues Associated with Worker Reentry Exposure Assessment*, FIFRA Scientific Advisory Panel, Washington, DC, December 2–5, 2008.
8. *EPA, A Set of Scientific Issues Being Considered by the Environmental Protection Agency Regarding: The Agency's Evaluation of Worker Reentry Exposure Assessment*, FIFRA Scientific Advisory Panel Meeting, SAP Minutes No. 2009-02, Washington, DC, December 2–4, 2008.
9. *EPA, Handler Exposure – EPA Presentation to the FIFRA Scientific Advisory Panel*, Washington, DC, October 6–8, 2009.

Chapter 5

Agricultural Reentry Task Force (ARTF)—History, Data Development, and Data Use

Michael E. Krolski*

ARTF Technical Committee Chair
Bayer CropScience, Chesterfield, Missouri 63017, United States

***Email: mike.krolski@bayer.com**

The United States Environmental Protection Agency (EPA) issued a Data Call-In (DCI) on October 18, 1995 requiring that pesticide manufacturers submit data on the persistence of chemical residue on crop foliage (termed Dislodgeable Foliar Residue or DFR) and on exposure to pesticides during reentry into any pesticide-treated agricultural site. The intention was for the data to be used to estimate exposure to agricultural workers when they enter fields that have been treated with a pesticide, and to calculate the restricted entry intervals that appear on product labeling. The Agricultural Reentry Task Force, L.L.C., (ARTF) was organized as a "joint data development task force" in which pesticide manufacturers worked together to develop a database that met EPA's requirements. ARTF worked closely with EPA, Canada's Pest Management Regulatory Agency (PMRA), the United States Department of Agriculture (USDA), and the California Department of Pesticide Regulation (CDPR) in developing the data and database. The Task Force had a Joint Regulatory Committee where representatives from these Agencies conferred with Task Force members on the regulatory direction of the project. ARTF's general plan for conducting a worker exposure study was approved by the Joint Regulatory Committee. ARTF data began to be used in government risk assessments when the database was put on-line in August 2000. This chapter will cover the establishment and structure of the ARTF, the design and performance of the studies completed to generate the necessary data, and the development of a database to allow regulators and registrants to access and use that data.

Introduction

Background

The United States Environmental Protection Agency (EPA) first published the requirements for protection of reentry workers in 1984 with the release of Subdivision K of the Pesticide Assessment Guidelines (*1*). Over time it became apparent that a standardized approach to determining worker reentry exposure was needed based on data from both foliar residue dissipation studies and potential exposure measurements by both dermal and inhalation routes. To that end, a Data Call-In (DCI) was published in 1995 requiring all pesticide manufacturers to submit data on the persistence of chemical residue on crop foliage and also on dermal and inhalation exposure to pesticides during reentry into any pesticide-treated agricultural site. The data were intended to be used to estimate exposure to agricultural workers when they enter fields that have been treated with a pesticide, and to calculate the restricted entry intervals (REIs) that appear on product labels.

In anticipation of that DCI, the Agricultural Reentry Task Force, L.L.C., (ARTF) was organized as a "joint data development task force" in which pesticide manufacturers worked together to develop an exposure database that met EPA's requirements. The ARTF is a limited-liability company formed by 31 crop protection companies in North America. These companies pooled their technical and financial resources to satisfy EPA's Data Call-In with the development of a comprehensive database that defines generic transfer coefficients (TCs) for agricultural reentry activities. To accomplish this, the member companies of ARTF collectively conducted a series of studies to generate data needed to form the database. The data from this research project may only be used to make regulatory decisions for the members who have sponsored the research or have compensated the ARTF for citation rights. Member companies of ARTF are:

ADAMA	Albaugh, LLC
Amvac Chemical Corporation	Arysta LifeScience NA
Atticus, LLC	BASF Corporation
Bayer CropScience	Dow AgroSciences, LLC
Drexel Chemical Company	DuPont Agriculture
FMC Corporation	Gharda Chemicals International, Inc.
ISK Biosciences	Loveland Products, Inc.
MEY Corporation	Mitsui Chemicals Agro, Inc.
Nissan Chemical America Corp.	Nufarm, Inc.
PBI-Gordon Corporation	Sharda Cropchem, Ltd.
Taminco US Inc.	Tessenderlo Kerley, Inc.
United Phosphorus, Inc.	Valent USA Corporation
Winfield Solutions, LLC	

To ensure that the regulatory end users of the data generated would get the data they needed in a timely and efficient manner, a Joint Regulatory Committee (JRC) was established to provide input and guidance to the ARTF starting from the planning stage and progressing all the way through database submission. JRC members included EPA, the United States Department of Agriculture (USDA), Canada's Pest Management Regulatory Agency (PMRA), and the California Department of Pesticide Regulation (CDPR). The JRC met regularly with ARTF to advise on the regulatory direction of the project. ARTF's general plan for conducting exposure studies was approved by the

JRC prior to implementation. Additionally, advice and input were obtained from the JRC on the conduct of each individual study before initiation ARTF data began to be used in government risk assessments when the database was put on-line in August 2000.

Risk

The risk of a toxicological response associated with the use of a chemical is a function of several variables:

1) The inherent toxicity of the compound
2) The magnitude of exposure to the compound
3) The route(s) of exposure to the compound
4) The ability of the system to detoxify the compound

In general terms, then, we can view worker exposure as being represented by the following equation:

$$\text{Risk} = f(\text{toxicity, exposure})$$

While the first and last of the variables listed previously are unique to each active ingredient, the magnitude and route of exposure to reentry workers can be considered to be related exclusively to the activity being performed and independent of the chemical being handled. Therefore, it is possible to generate a set of universal or "generic" exposure values that can be consistently applied to all agricultural chemicals for a given activity. This principal allowed the ARTF to develop data based on a limited set of representative registered active ingredients that served as surrogates for all actives.

One corollary of this approach to risk is the realization that in the absence of exposure there is no risk. An example of this situation is the harvesting of crops using closed cab equipment (e.g. combines, cotton pickers, sugarcane harvesters), where the operator does not come into physical contact with the treated crop. For this particular activity, the mitigation of exposure through the engineering control of the closed cab overcomes any potential toxicological issues.

ARTF Data Collection Activities

Transfer Coefficients

Pesticide residues will "transfer" from a crop to a worker primarily when contact is made between the worker and the crop. There are several variables to consider that may influence this transfer, including the crop, worker activities, plant height, and foliage type and density. ARTF could not test all crop, activity, and growth stage combinations, so a set of "representative" and important combinations were tested. Then, one of the tested combinations was assigned to surrogate for other, similar, possible combinations. The basis for the translation was the development of a set of standardized transfer coefficients.

A TC is the proportion of available residue on a crop that is transferred to a worker who comes into contact with that crop. Determination of the TC is dependent on knowing how much residue is available for transfer at the time of contact, so a concurrent Dislodgeable Foliar Residue (DFR) study must be performed when determining a TC value. DFR values were found to be dependent on leaf texture. Initial DFR from waxy leaves are generally high because there is nothing holding the residues on the foliage compared to hairy leaves where residues are difficult to remove due to the large amount of pubescence. In general, DFRs over time were log normally distributed. Across the entire measured dataset initial DFR accounted for approximately 25% of the applied active ingredient and DFRs dissipated at a rate of approximately 10% per day. Once known, an established TC can then be combined with a product specific measured DFR value to estimate exposure to the active ingredient of interest. The TC value for a given activity is determined by dividing the measured exposure of a worker over a unit of time by the amount of residue available on the leaf surface for transfer, as shown as follows:

TC = Measured exposure during time worked/Available Residues on crop

$$= (\mu g/hr)/(\mu g/cm^2)$$

$$= cm^2/hr$$

The units for transfer coefficients (cm^2/hr) may seem strange. They can be viewed conceptionally as a measure of the intensity of contact between the worker and the foliage for a given crop and activity combination.

Surveys and Purchased Studies

Reentry activities can be combined into groups of crop/activities that involve similar contact with crop foliage (and thus similar exposure and transfer coefficients). The ARTF conducted a Growers Survey to help identify which crop/activity combinations would have similar exposure. Growers and grower consultants were asked to estimate the relative amount of contact between crop foliage and various body parts of farm workers, in terms of frequency and degree of contact. The survey produced information on approximately 3000 crop and activity combinations in North America. ARTF, with the JRC, used the survey data to determine the crops and activities on which exposure studies would be most informative. The data were also used to categorize the various crops and activities within the ARTF database.

To minimize duplication of effort, ARTF purchased 18 studies that were conducted by member companies prior to the establishment of the Task Force. These studies comprised 283 separate sets of exposure measurements, each set quantitatively describing exposure to the various body parts of one worker entering a field for one workday to perform one activity in one crop. All of the studies were reviewed for acceptability by independent and regulatory reviewers prior to becoming part of the ARTF database. Some additional studies brought to ARTF for consideration were not included in the database. Only those sets of exposure measurements that the reviewers deemed scientifically acceptable were purchased. Applicable data were thoroughly reviewed and entered into the ARTF database.

Clustering of Worker Reentry Activities

In consultation with the JRC, reentry activities identified in the survey process were divided into ten groups:

1. No Foliar Contact
2. Hand Irrigation – all crops

3. Intense Contact – all crops
4. Field Crops, smooth leaves
5. Field Crops, hairy leaves
6. Field Crops, waxy leaves
7. Trellis Crops
8. Orchard Crops
9. Grass and Sod
10. Nursery and Greenhouse (including transplanting)

Of the 3000 crop/activity/height/foliage combinations identified, 1400 were determined to be in group 1, with no worker contact, and so no TC was needed. The remaining crop/activity/height/foliage combinations were divided into clusters of similar activities, and the proposed clustering was reviewed by an outside panel of academic and agronomic experts. The final set of 28 clusters are identified in Table 1.

Studies Performed by ARTF

General

Having identified the crop/activity/height/foliage clusters that required additional data generation, the ARTF began planning studies. It was decided to develop a standardized format for all study protocols and use a common set of standard operating procedures (SOPs) and a standard Informed Consent Form. Additionally, a standardized protocol was developed for use in all of the concurrent DFR studies to be performed with the worker exposure studies. All of these activities were approved by the JRC prior to the start of data collection by ARTF.

ARTF conducted 29 field studies, comprising 482 replicates (worker-days). Typically, each study included two applications of the surrogate, applied with a 7-day retreatment interval, at normal application rates for the chosen surrogate chemical. Multiple reentry workers (typically 5 per time point) were then monitored performing a day's work on three different days within a week following the last application. Concurrently, DFR samples were collected on each day worker monitoring was performed to provide an accurate measure of residues available for transfer.

Surrogate Chemicals

It was also decided to use a limited set of surrogate chemicals for these studies. All of the exposure studies used test substances that were already registered pesticides. The surrogate compounds selected had the following desirable characteristics:

- Registered for use on a variety of crops.
- Relatively short Restricted Entry Intervals (REIs).
- Short Preharvest Intervals (PHIs) for harvest studies.
- Known to be stable under field conditions.
- Relatively persistent residues on foliage.
- Validated analytical methods that were practical, reproducible, and sufficiently sensitive for detection of the test substance in all types of experimental matrices.

Table 1. ARTF Cluster Codes and Cluster Descriptions

Cluster Code	*Cluster Description*
Sx	Smooth-leaf field crops: intense contact activities
SSs	Smooth-leaf field crops: scouting in solid stand conditions
SSr	Smooth-leaf field crops: scouting in row conditions
SH	Smooth-leaf field crops: hand harvesting and tying
SW	Smooth-leaf field crops: hand weeding, thinning, and similar contact activities
HH	Hairy-leaf field crops: hand harvesting and similar contact activities
HHt	Hairy-leaf Tobacco: hand harvesting and canopy management
HS	Hairy-leaf field crops: scouting and similar contact activities
Wm	Waxy-leaf field crops, medium height: all activities plus full foliage weeding
WlH	Waxy-leaf field crops, low height: hand harvesting and similar activities
WlS	Waxy-leaf field crops, low height: scouting and similar contact activities
Tx	Trellis crops: intense contact activities
THg	Trellis crops: hand harvesting juice/wine grapes and similar contact activities
THb	Trellis crops: hand harvesting caneberries and similar contact activities
TP	Trellis crops: hand pruning, scouting, and similar contact activities
OH	Orchard crops: hand harvesting and similar contact activities
OHn	Orchard crops: mechanically harvesting nuts
OP	Orchard crops: hand pruning, scouting, and similar contact activities
OW	Orchard crops: hand weeding and similar contact activities
DM	Golf courses: maintenance activities
DH	Sod: mechanical harvesting, scouting, transplanting, and hand weeding
GHf	Greenhouse floriculture hand harvesting: all flowers and methods
GHv	Greenhouse vegetables: hand harvesting and similar contact activities
GN	Greenhouse and nursery crops (others): all activities
I	Irrigation, any crop where hand line is possible
CHp	Cotton, mechanical harvesting: picker operator and raker
CHm	Cotton, mechanical harvesting: module builder operator
CHt	Cotton, mechanical harvesting: tramper

The process of approving a compound as a surrogate involved several steps:

1. A potential surrogate was identified.
2. Analytical methods were either developed (if none existed) or modified to accommodate the sensitivity required by ARTF studies.

3. The analytical methods were validated in a study conducted under EPA Good Laboratory Practice Standards.
4. The stability of the surrogate under field conditions was evaluated within a small field study.
5. Freezer storage stability of the potential surrogate was also evaluated.

Once a compound passed all of these tests and was accepted as a surrogate, the ARTF Analytical Subcommittee selected analytical laboratories to perform a try-out of the analytical methods. Laboratories that were successful were then considered eligible to analyze samples from ARTF field studies.

The ARTF selected the following compounds as surrogate chemicals for generic measurement of exposure: chlorothalonil, malathion, diazinon, carbaryl, and cyfluthrin.

Study Participants and Sample Collection

All ARTF studies were performed prior to adoption of the Human Studies Rule (2) by EPA. All studies were performed under a protocol approved by the ARTF Technical Committee, JRC, and an Independent Review Board (IRB) for ethical considerations. Workers were recruited by the study director, who advised the volunteer participants on the nature and requirements of the study and obtained written informed consent for each participant.

Worker Exposure

In the ARTF worker reentry exposure studies, workers were dressed in standard clothing while they performed their normal activities after a field was sprayed with a surrogate compound. At the end of the day, the amount of dermal exposure was determined by measuring the amount of surrogate residue on the inner and outer clothing, the face/neck area, and hands. Inhalation exposure was determined by measuring the amount of surrogate collected on air sampling tubes attached all day to an air pump. Sample descriptions and collection procedures are outlined as described in the following sections.

Prior to monitoring each day, workers were dressed in two layers of 100% cotton clothing (inner and outer dosimeters). The inner layer consisted of 100% cotton long underwear that had been double-rinsed with methanol and the outer layer was a 100% cotton white long-sleeved shirt and long pants, which were provided by the ARTF. This clothing was worn throughout the period of monitoring (the working day) and removed at the end, with the assistance of a member of the monitoring team. The inner layer of clothing (inner dosimeter) was sectioned into upper and lower arms, front and back torso, and upper and lower legs, and placed into temporary frozen storage as soon as possible for transport to the analytical facility. The outer layer of clothing (outer dosimeter) was typically also sectioned and stored frozen for determination of total dermal deposition (TDD), if those data were desired. Samples were shipped frozen to the analytical lab for assay.

Hand exposure was assessed by having the workers wash their hands in a 0.01% v/v aqueous Aerosol OT solution (AOT; sodium dioctyl sulfosuccinate, a mild detergent) according to a standardized washing procedure. The hand wash process consisted of the worker scrubbing their hands over a clean bowl for 30 seconds while a researcher poured 250 mL of the wash solution over their hands. The procedure was repeated and the solutions were combined for a single 500 mL sample for each hand wash event. A hand wash was performed just prior to beginning the work activity and that hand wash was discarded. Hand wash samples were collected each day at the end of each exposure monitoring period, and at any point when a worker would normally wash their hands (i.e., prior to eating or using the toilet). Hand washes collected during and at the end of the work period were each treated as separate samples. Hand wash solutions were stored and shipped frozen to the analytical lab for assay.

Face and neck exposure were measured by sequentially wiping the exposed areas with two gauze pads that had been wetted with the same 0.01% v/v AOT solution used for hand wash samples. The face and neck area were wiped each sampling day just prior to entering the field. These wipe samples were discarded. Samples were collected prior to eating and again at the end of the work period. Samples collected during a single workday were stored frozen and later combined for each worker with a final wipe, resulting in a single sample for analysis.

Airborne concentration of active ingredients in the worker's breathing zone were monitored with an air sampler connected by Tygon® tubing to a personal air sampling pump. The pumps were calibrated before each workday to operate at a flow rate of approximately 2.0 liters per minute (LPM). The sampler was held in a plastic tube holder and clipped to the worker's outer shirt collar with the intake facing downward, placing the collection tube in the worker's breathing zone. The air sampling pumps operated throughout the monitoring period including breaks and lunch. At the end of each work day the Tygon® tube, pump and sampler were removed from the worker and the airflow rate was measured. The sampler was then disconnected from the tubing, capped at both ends and placed in a freezer for shipment to the analytical lab.

Dislodgeable Foliar Residue Sampling

Associated with each worker exposure study, a concurrent DFR study was performed. Separate control and treated plots were established, with the treated plot within the same treated area used for worker exposure but used exclusively for DFR sampling. Samples were collected using a Birkestrand leaf punch sampler with a 2.54 centimeter, or 1 inch, diameter punch. The leaves were sampled directly into pre-labeled glass jars with each sample consisting of 40 leaf disks for a total of approximately 400 cm^2 leaf surface area (counting both sides of the foliage). At each time interval, a single sample was collected from the control plot and triplicate samples were collected from the treated plot. To establish dissipation curves, samples were collected pretreatment, on the day of the application (4 hours after the second application or when the spray was dry, whichever was the latest), and 1, 2, 3, 4, 5, 6, 7, 10, 14, 21 and 28 days following the second application,

Residues were dislodged from the leaf punch samples with 0.01% Aerosol OT solution. Leaf disks were dislodged within the sampling jars, twice, for 10 minutes each time, with 100 mL of the Aerosol OT solution on a reciprocating shaker operating at approximately 200 cycles per minute. The dislodging solutions were combined for each sample and frozen for shipment to the analytical lab.

Calculation of Dermal Exposure and TCs

Raw residue values for each dermal exposure matrix were determined using previously validated analytical methods. These raw residue values were adjusted based on the average recoveries from field fortification samples generated along with each sampling event in a given study. There was no adjustment for the differences between face/neck wipe measurements and total head area. Two different dermal exposure values were calculated: potential dermal exposure (PDE) was the sum of all dermal exposure measurements, and total dermal exposure (TDE) was the sum of measurements except for outer dosimeters. Conceptually, PDE is the amount of residue that may be transferred to the worker while TDE is the amount of residue that may actually contact the skin and be available for absorption.

Having TDE values and the associated DFR values at the time of sampling, TC values could be calculated for each sampling interval:

$$TC = [(TDE)/Exposure\ Time]/DFR$$

The mean TC was then calculated by averaging the three total dermal TCs determined from measurements made on each study day. The mean TC is time independent and can thus be used to calculate potential exposure at any time where an estimate of DFR is available.

The resulting TC values determined experimentally for each study were then assigned as surrogate values for similar crop/activity/height/foliage activities as outlined in the clustering scheme.

Database Review and Adoption

In an effort to confirm its scientific approach, ARTF sponsored a peer review of its methods. In October 2000, ten experts in areas such as, exposure, agronomy, and statistics, convened to review and discuss the ARTF approach and methods and how ARTF had responded to EPA's Data Call-In.

ARTF received a positive response from the peer reviewers, along with some suggestions, to confirm ARTF had developed a scientifically sound database of generic transfer coefficients to use for risk assessments. In response, ARTF performed several statistical analyses to confirm the cluster assignments of the various crop/activities, and to better define the confidence limits on mean TCs. The reviewers concluded that the ARTF database framework was very appropriate for risk assessments and the data were of significantly higher quality than that on which regulatory default values was based at that time.

All of the clustering decisions are summarized, and a final list of crop activity assignments provided, in the 2006 revised ARTF TC Database Report (3). The final ARTF TC Database contains information from 47 reentry exposure studies, with a full set of body-part TCs for over 750 worker-days. These data are organized into 28 TC clusters that have a similar magnitude and distribution of exposure potential among body parts. The database also contains over 3,000 crop/activity/height/foliage combinations assigned to those clusters to be represented by cluster-specific TC data during risk assessments. In addition, the database identifies approximately 1,400 crop/activity/height/foliage combinations that involve little or no contact between reentry workers and foliage and for which the use of a TC is not appropriate.

The overall ARTF program was submitted to USEPA which, in turn, established a Science Advisory Panel (SAP) in 2008 to independently review the study conduct, clustering proposal, survey methodology and results, and peer review process. Ultimately, the SAP was asked to evaluate if the ARTF program fulfilled the requirements outlined in the 1995 DCI.

The SAP was empaneled and met in December 2008 chaired by Janice Chambers from the College of Veterinary Medicine at Mississippi State University. The ARTF made a presentation that highlighted the overall background, results of the expert and grower surveys, and the clustering process. Another presentation described the conservative nature of the ARTF data when compared to biological monitoring results using data from previous apple thinning and apple harvesting studies.

The overall response from the SAP was very positive (4). The ARTF was complimented on the volume of data from the grower survey and from the exposure studies; and the quality of the presentations. The SAP determined that ARTF data represent the best and most extensive data available.

The ARTF clustering scheme was considered logical with only a few, relatively minor, improvements suggested. The amount of data generated by the grower survey was considered sufficient for identifying crops and activities. The SAP particularly liked the approach of basing the clustering more on the agronomic and scientific factors rather than on statistics. The data are highly variable, which is typical of exposure data, but the ARTF has done the best possible job of dealing with the data sets. It was noted that there are often only small differences between clusters, indicating it might be possible to collapse the present cluster into a smaller number of "super clusters".

Conclusion

ARTF was formed in anticipation of a DCI for reentry worker exposure data, under the premise that through standardization and cost sharing, Task Force member companies could provide the required data better and faster than as individual companies. ARTF was able to generate a common database for use by both regulators and registrants for evaluating the safety of crop protection products.

By working with regulators through the JRC from its inception, ARTF provided the information that was desired by regulators in a timely and efficient manner, maximizing the resources of both regulators and the regulated community. The organizational processes and experimental procedures developed by the ARTF have been applied by subsequent task forces, proving the validity of this approach.

Acknowledgments

The author would like to thank the 31 Member Companies for their participation and support and especially the individual member company representatives for their efforts to develop the ARTF program and database. Thank you to the JRC members who ensured both the integrity and quality of the work. Finally, a special thanks goes to the ARTF Task Force Manager, Dr. Dave Johnson; administrative assistant, Rhonda Bischel; and two Task Force consultants who consistently went above and beyond in their efforts to generate a quality database, Eric Bruce and Dr. Dennis Klonne.

References

1. Pesticide Assessment Guidelines Subdivision K, Exposure: Reentry Protection. 40 CFR 158.140, 5 October, 1984; US Environmental Protection Agency, U.S. Government Printing Office: Washington, DC, 1984.
2. Protections for Subjects in Human Research; US Environmental Protection Agency, U.S. Government Printing Office: Washington, DC, 2006.
3. Bruce, E.; Holden, L.; Korpalski, S. Reissued Report: Agricultural Reentry Task Force Transfer Coefficient Database: A Summary of its Development and Use; ARTF LLC: Macon, MO, January 2006. (MRID 46734002).
4. SAP Minutes No. 2009-02, Memorandum from S. R. Matten to D. Edwards Dated 19 February, 2009; US Environmental Protection Agency, U.S. Government Printing Office: Washington, DC, 2009.

Chapter 6

Overview of the Outdoor Residential Exposure Task Force (ORETF)

Jennifer M. Thomasen*

ORETF Technical Committee Chair, Bayer Crop Science, Chesterfield, Missouri 63017, United States

***Email: jennifer.thomasen@bayer.com**

The Outdoor Residential Exposure Task Force (ORETF) was formed in 1994 in anticipation of a United States Environmental Protection Agency (USEPA) Data Call-In (DCI). The USEPA issued a DCI on March 3, 1995 requiring that pesticide manufacturers submit exposure data on the use of pesticides on residential grass, lawn, or turf settings. The ORETF was organized as a "joint data development task force" in which pesticide manufacturers worked together to develop data that met regulatory requirements. The ORETF worked closely with USEPA, Canada's Pest Management Regulatory Agency (PMRA), the United States Department of Agriculture (USDA), and the California Department of Pesticide Regulation (CDPR) in developing the data. Collectively, these regulatory agencies formed the Joint Regulatory Committee (JRC). The ORETF representatives met with the JRC throughout the data development for advisement on the regulatory direction of the program. The ORETF first developed a standard technique for measuring turf transferable residues and then developed a procedure for defining activities representative of children playing on turf. A comprehensive reentry study was conducted in 2002 and the results submitted in 2004. The data and methods developed by the ORETF are currently used for regulatory risk assessment. The ORETF terminated its existence as a limited liability company (LLC) in July 2020.

Introduction

The United States Environmental Protection Agency (USEPA) requires, pursuant with the Federal Insecticide, Fungicide, and Rodenticide Act (FIFRA), pesticide registrants to submit exposure data (i.e., how much pesticide residue contacts the clothing and skin of individuals) for the purpose of conducting risk assessments. The USEPA was aware of widespread use of pesticides at residential use sites. This caused increasing concern about the potential risk to both professionals and non-professionals (residents) from exposure to pesticides at these residential use sites. A Data Call-In (DCI) was published in 1995 requiring all pesticide registrants of products (herbicide, fungicides, insecticides, rodenticides) used on residential grass, lawn, or turf settings to submit data to be used in estimating risk at these sites. The DCI specifically identified the need for exposure data of homeowners and professional lawn care operators during pesticide applications as well as exposure to individuals who enter a residential turf area following pesticide application for 117 active ingredients. Table 1 provides a list of the data required by the USEPA DCI notice.

Table 1. Summary of USEPA DCI

Guideline Number	*Study Title*
231 (875.11)	Dermal Exposure to Mixers, Loaders and Applicators at Outdoor Sites
232 (875.13)	Inhalation Exposure to Mixers, Loaders and Applicators at Outdoor Sites
132-1(a) (875.21)	Foliar Residue Dissipation
133-3 (875.25)	Dermal Exposure Upon Reentry to a Treated Area
133-4 (875.25)	Inhalation Exposure Upon Reentry – Agency Reversed

Formation of the Outdoor Residential Exposure Task Force (ORETF)

In anticipation of the DCI, the Outdoor Residential Exposure Task Force, LLC., (ORETF) was organized as a "joint data development task force" under FIFRA. This allowed pesticide manufacturers to work together to develop exposure data that met USEPA's requirements. USEPA announced the formation of ORETF on December 9, 1994 in PR Notice 94-9. The ORETF is a limited liability company formed by 33 crop protection companies in North America. The formation of the ORETF allowed for member companies to share scientific expertise and financial resources to develop a more consistent and comprehensive exposure database than could be developed by any individual member company. The formation of the ORETF also allowed member companies to work as a single entity with regulatory experts to improve efficiencies and ensure the data developed would meet regulatory needs.

To accomplish this, the member companies of ORETF collectively conducted research to generate data to evaluate residential exposure. The member companies pooled their technical and financial resources to satisfy USEPA's DCI and developed a comprehensive database defining the generic relationship between exposure and turf transferable residues (TTR) and generated exposure data for professional lawn care operators and homeowners who mix and apply pesticides to residential lawns, gardens and ornamentals. The data from this research project may only be used to make regulatory decisions for the members who have sponsored the research or have compensated the ORETF for citation rights. Member companies of the ORETF are:

ADAMA
Albaugh, LLC
Amvac Chemical Corporation
Arysta LifeScience NA
Atticus, LLC
BASF Corporation
Bayer Crop Science
Dow AgroSciences, LLC
Drexel Chemical Company
DuPont Agriculture
Ensystex, Inc.
FMC Corporation
Gharda Chemicals International, Inc.
Gowan Company
Helena Agri-Enterprises, LLC
ISK Biosciences
LG Chem, Ltd.
Loveland Products, Inc.
Luxembourg-Pamol, Inc.
McLaughlin Gormley King Co

MEY Corporation
Mitsui Chemicals Agro, Inc.
Monsanto Company
Nippon Soda Co., LTD.
Nufarm, Inc.
PBI-Gordon Corporation
Sharda Cropchem, Ltd.
Sipcam Agro USA, Inc.
Syngenta Crop Protection
The Scotts Company
United Phosphorus, Inc.
Valent USA Corporation
Winfield Solutions, LLC

To ensure that the regulatory end users of the data generated would get the data they needed in a timely and efficient manner, a Joint Regulatory Committee (JRC) was established to provide input and guidance to the ORETF starting from the planning stage and progressing all the way through data submission. The JRC members included USEPA, the U. S. Department of Agriculture (USDA), Canada's Pest Management Regulatory Agency (PMRA), and the California Department of Pesticide Regulation (CDPR). The JRC met regularly with the ORETF to advise on the regulatory direction of the project. The ORETF's general plan for conducting exposure studies was approved by the JRC prior to implementation.

The passage of the Food Quality Protection Act (FQPA) in 1996 further increased the importance of estimating residential exposure to pesticides because it required multi-route (oral, dermal, and inhalation) and multi-source (food, water, residential, pet care) aggregation of exposure and risk. This led the ORETF to further expand its original scope. The ORETF data generated is currently being used in government risk assessments and has been incorporated into USEPA Standard Operating Procedures for Residential Pesticide Exposure Assessment published October 2012 (*1*).

ORETF Program

Generic Exposure Principle

The ORETF program was based on the concept of "generic exposure" which was possible because previous research established that pesticide exposures are primarily influenced by human activities involved with handling or contacting the products and not the chemical nature of the active ingredient in the product. Exposure is not determined by the properties of the active ingredient in a product but rather by physical factors and activities involved in handling the product. This allows for the development of generic exposure databases, where exposure is generated from a specific set of surrogate active ingredients that can be used to predict exposure occurring from any active ingredient in a product that is used under similar conditions. The generic data principle can be applied to handler as well as post-application exposure and has been accepted since the 1980s (2). For example, the amount of chemical residue that contacts an individual's clothing and skin is a function of the formulation type (i.e., liquid or granule product), the way in which a person handles the pesticide when mixing and applying product, and the degree of contact between the body and treated turf/ foliage when reentering a treated area. The specific active ingredient in the pesticide product does not affect the magnitude of exposure.

The overall ORETF program consisted of the development of generic exposure data to evaluate human exposures resulting from pesticide use on residential turf. Studies were conducted to monitor handler exposure (those that mix, load, and/or apply) for lawns, gardens, and ornamentals as well as post-application or reentry exposure to treated turf.

Handler Exposure

A pesticide handler is someone who mixes, loads, and/or applies pesticides. The ORETF conducted mixer, loader, applicator (M/L/A) studies that determined exposure to homeowners and lawn care operators applying liquid and granular products. Potential dermal exposure was monitored using inner and outer whole body dosimeters, together with hand washes and face/neck wipes. Potential inhalation exposure was monitored with personal sampling pumps. The same procedures were used across all the studies.

The ORETF generated generic exposure data for the following outdoor residential scenarios summarized in Table 2.

Table 2. Summary of ORETF Handler Exposure Data Scenarios

Population	*Activity*	*Target Area*	*Formulation*
Professional Lawn Care Operators (LCO)	Mix/Load/Apply	Lawns	Liquid
			Granule
Homeowner	Mix/Load/Apply	Lawns	Liquid
			Granule
		Ornamental Plants	Liquid
		Garden Vegetables	Liquid
			Dust

Briefly, outer and inner dosimeters consisted of 100% white, cotton material. The outer dosimeter included a long-sleeved shirt and long pants, while the inner dosimeter consisted of long underwear. Hands were washed with a 0.01% Aerosol® OT (AOT) solution (sodium dioctyl sulfosuccinate in distilled water). AOT-moistened gauze pads were used for the face/neck wipes. Airborne concentrations were monitored with a personal low-volume air sampling pump with an appropriate sample sorbent proceeded by a glass-fiber filter and contained in an OVS tube (OSHA Versatile Sampler).

Field fortified samples (generally in duplicate or triplicate) were prepared for each matrix on each day of monitoring using the appropriate analytical standard. All analytical methods used for the study matrices were validated prior to use and met the USEPA recommended guidelines for recoveries. Appropriate standards, solvent blanks, lab controls, and concurrent lab spikes were analyzed with each set of samples for each matrix.

All ORETF mixer, loader and applicator studies were conducted according to established regulatory guidelines for exposure studies and the Good Laboratory Practice (GLP) regulations. These studies were completed and submitted to regulatory authorities in 2002 (3). These studies also met the requirements of CDPR and PMRA.

Post-application Exposure

A major challenge when evaluating residential post-application is exposure is that, unlike a mixer/loader or applicator, post-application is not a single, easily defined activity. Therefore, in order to conduct a residential post-application exposure assessment, it is necessary to define a standardized set of activities considered representative of the types of activities that may occur on residential turf. The ORETF developed a methodology for determining potential exposure of children and adults reentering lawns treated with pesticides, post-application exposure. Based on the generic characteristic of exposure, the magnitude of pesticide residue contacting an individual is not a function of the active ingredient. The amount of pesticide residue contacting an individual reentering lawns treated with pesticides is a function of the formulation type (liquid or granule) and the amount of contact with the treated surface.

To determine potential exposure of children and adults reentering lawns treated with pesticides, a standard methodology was needed to measure the residues on the turf or the turf transferable residues (TTR). Along with the TTR, a measure of the residue on the person is needed to determine the degree of contact that is specific to the activity. This allows for the calculation of a transfer coefficient (TC). A detailed explanation of TCs is provided in the chapter by Krolski.

$$TC = \text{measured exposure during specific activity/TTR}$$

To fully understand post-application exposure, the ORETF needed to develop techniques to measure the TTR and determine the representative post-application activities on residential turf. To achieve this objective, ORETF:

1. Developed a standardized method to quantify turf transferable residues (TTR).
2. Analyzed use and activity pattern survey data to better understand residential pesticide use patterns and post-application behavior patterns.
3. Analyzed Jazzercise® and videography data to quantify the nature of dermal exposure to pesticides for adults and children reentering treated turf.
4. Developed the Children's Activity Pattern (CHAPS) routine using activity pattern survey data and videography data.
5. Conducted a post-application exposure study to generate data on amount of pesticide residue transferred to individuals who enter turf immediately following pesticide application.

The next sections will briefly describe the work conducted to complete 1-5 and ultimately achieve ORETF's objective.

Measure Turf Transferable Residues

Conducting exposure assessment for people reentering turf that has been treated with a pesticide requires an assessment of the amount of residue that can be transferred from the grass to the person. Several techniques had been proposed for assessing the amount of turf transferable residues (TTR). The goal of the ORETF was to establish a standardized method for measuring turf transferable residues that was reproducible within and between individuals conducting the sampling on different days, different turf types, different geographies, etc.

The ORETF investigated five established turf sampling techniques for high sensitivity and low variability among and between individuals (*4*). From these evaluations, the ORETF developed a single standardized technique for measuring product-specific turf transferable residues. The technique, named the Modified California Roller technique, consists of pushing a 32-pound roller 10 times across a 6-square-foot area of percale cloth that is securely fastened to the ground on top of the turf. The Modified California Roller methodology has been published (*5, 6*) and incorporated into the USEPA guidelines for exposure studies (875.2100).

A. Jazzercise®

Jazzercise® is an aerobic exercise routine developed by Judi Sheppard Missette that was very popular throughout the 1990s. The Jazzercise® routine is choreographed and a highly reproducible set of activities. Jazzercise® was used to evaluate exposure to indoor surfaces and later turf because it is a standard activity routine involving a high degree of contact with the treated surface. Prior to the formation of the ORETF, the relationship between exposure and turf transferable residue was determined by monitoring the amount of residue that contacted the clothing and skin of individuals who performed a standardized 20-minute Jazzercise® routine on pesticide treated turf. Therefore, use of the Jazzercise® routine was intended to be a reproducible and high-end estimate of the amount of contact with a treated surface as a measure of human exposure under variable environmental conditions.

By holding contact constant through the use of Jazzercise®, the method permits investigation of a variety of parameters. These include the effect of different pesticides, different reentry periods, variable amounts of clothing, aging surface residues, variations of temperature and humidity, effects of sequential exposure to the same treated area or new treated areas, etc. Jazzercise® was not designed to mimic children's (or adults') behavior but it does provide a reproducible estimate of exposure under known conditions that can be related to children's activities. Actual human activity on treated turf would likely result in a lower degree of contact with the turf, and thus a lower exposure than the Jazzercise® routine.

The ORETF proposed that a 20-minute Jazzercise® routine could be used conservatively to estimate the daily contact of a child to residential turf. The ORETF hypothesized that adult aerobic/floor exercise could be related to children's activities by using videography data developed at Stanford University on both children's activity patterns and the Jazzercise® routine. The ORETF sponsored a study at Stanford University using video analysis of children playing on turf to estimate surface contact duration and frequency for 14 regions of the body in order to gain an understanding of the micro-level activities of children. The same micro-level video analysis technique was applied to an adult performing a standard Jazzercise® routine. Exposure data that were normalized for variables such as frequency (contacts per hour), contact length (seconds per contact) and the resulting contact duration (seconds per hour) in a Jazzercise® study were then compared to children's contact rates on residential turf. The data from this adult-to-child comparison supported the conclusion that 20 minutes of an adult Jazzercise® routine provides a conservative basis for estimating children's daily exposure to residential turf. The Stanford University videography not only allowed for the comparison in the degree of contact between children's activities and the Jazzercise® routine, but also the degree of hand-to-mouth activity for children was determined (*7, 8*).

The ORETF proposed using existing Jazzercise® data plus the difference in degree of contact between children versus Jazzercise® determined from the Stanford University videography study to estimate potential dermal exposure to children reentering lawns treated with pesticides. However, this approach was criticized as the Jazzercise® routine did not in any way simulate the actual activities that individuals perform on turf and uncertainty remained regarding whether Jazzercise® was a realistic measure of post-application exposure. Further, concerns regarding existing Jazzercise®

studies arose because they were conducted using various TTR techniques. The TTRs in these existing study data were a higher percentage of the application rate than the newly conducted ORETF TTR studies using the Modified California Roller. Following a JRC meeting it was decided that the ORETF would develop a more realistic and representative routine of activities commonly performed by children on residential turf .

B. Children's Activity Patterns (CHAPS)

The ORETF designed the Children's Activity Patterns (CHAPS) routine to reflect the types of activities commonly performed by children on residential turf. The ORETF used data from a National Human Activity Patterns Survey (NHAPS), conducted by USEPA, to identify the specific activities performed by children playing on turf and the duration of time spent performing each activity (9). With the NHAPS data and the Stanford University videography data, the ORETF developed the CHAPS routine. The CHAPS routine was designed so the degree of contact with the turf equaled or exceeded the degree of contact measured in the Stanford University videography study and NHAPS. CHAPS is a 2-hour routine in which adults performed activities commonly performed by children on residential turf.

Briefly, the 2-hour CHAPS routine consisted of 3 activity types for a percentage of time: passive (50%), active (35%), and hard direct (15%). The activities in the CHAPS routine were as follows:

- Passive (5 activities x 12 minutes: walking/jogging, playing catch, crawling, picnicking, resting)
- Active (5 activities x 8 minutes: playing with toys, playing soccer, playing frisbee, playing games, playing tag)
- Hard direct (2 activities x 10 minutes: tumbling, football)

Post-application Exposure Study

On April 10-11, 2002 the ORETF conducted a post-application exposure study at Grant County Fairgrounds in Moses Lake, Washington. This study required 100 people on site for 2 days. The surrogate chemical used for this study was Dithiopyr. This was selected because it was a commonly used residential pesticide product. The product was also available in both liquid and granular formulations and was applied at a rate of 0.5 lb. of active ingredient per acre.

The objective of this study was to generate data on the amount of pesticide product residue transferred to individuals who enter turf immediately following pesticide application. This data was necessary to allow the ORETF to develop an algorithm for determining post-application reentry exposure for adults and children that would be used by pesticide registrants and regulatory agencies. The study was designed to allow for the comparison of the Jazzercise® and CHAPS routines. To address concerns regarding the higher TTRs in the existing Jazzercise® studies compared to the more recent ORETF TTR data, a low application rate and low TTR were utilized in this study. This lower rate was also reflective of the decreasing application rates for many products on the market

On the first day of the study, a granular application was made to three-inch bluegrass turf. Monitoring was performed on 10 men and 10 women performing the 20-minute Jazzercise® routine. Monitoring was also performed on another group of 10 men and 10 women performing the 2-hour CHAPS routine. On the second day, a liquid application was made to three-inch bluegrass turf. The same monitoring and activities were repeated as the first day. Routines were performed by local adult volunteers.

Volunteers wore outer and inner dosimeters consisting of 100% white, cotton material. The outer dosimeter was a coverall, while the inner dosimeter consisted of long underwear. The cloth dosimeters were cut into six body parts for analysis to allow for the calculation of exposure for various clothing scenarios. Face/neck wipes and hand washes were collected as previously described. Additionally, socks on the feet and foot washes were collected The data could be combined from the various sections of inner and outer dosimeters to provide exposure for several types of clothing scenarios such as the standard long-sleeved shirt and long pants with shoes and socks; shorts and t-shirts; and no clothes. These different clothing scenarios are applicable to residential turf exposure.

During the study, TTR samples were collected using the Modified California Roller TTR technique. Five samples were collected over the duration of each routine and were evenly spaced. Field fortification samples were collected for all matrices to allow for the adjustment of field sample values if necessary.

The objectives of the post-application study were achieved. The ORETF successfully completed exposure evaluations for the CHAPS and Jazzercise® routines and determined the relationship between the two. The study achieved low TTRs which made it possible to determine the relationship of data from these Jazzercise® studies to previous Jazzercise® studies with high TTRs. On a formulation basis, when normalized for time, the Jazzercise® and CHAPS routines resulted in very similar dermal loadings. Point transfer coefficient (TC) estimates for post-application dermal assessment to lawns/turf were derived. The algorithm developed from this study provides a more refined and realistic means of determining the relationship between TTR and exposure (*10*). This data is incorporated into the 2012 USEPA Standard Operating Procedures for Residential Pesticide Exposure Assessment (*1*).

Additional Contributions to Residential Exposure and Risk Assessment

Hand-to-Mouth Exposure

The ORETF gained a better understand oral exposure potential resulting from hand-to-mouth activity by using the videos from the Stanford University videography study to record the frequency of hand-to-mouth events and the surface area of the hand that was mouthed per event. The ORETF submitted two reports to the regulatory agencies that provide refined and more realistic data for hand-to-mouth exposure (*7, 8*). This research assisted to facilitate the development of risk assessment tools and further the understanding of incidental oral exposure potential resulting from hand-to-mouth activities.

Surveys on Pesticide Usage and Lawn Sizes

ORETF generated extensive pesticide use and usage data from two surveys – one sponsored by ORETF and one that was an extension of an annual survey conducted by the National Gardening Association (*11*). These survey data provide realistic use data related to exposure (frequency, time, amount) that can be used for exposure and risk assessment.

The ORETF-sponsored Use and Usage Survey was conducted to establish a database of information on the use of pesticides in outdoor residential settings. The ORETF Use and Usage Survey consisted of phone interviews in June, August, and October of 1998 to obtain information regarding the mixing, loading, and application of home pesticides in the United States and Canada. At each of these three times (i.e., three waves), respondents were asked about their pesticide use in the previous two months. Phone interviews were conducted in 150 households in each of six geographic regions for each of the three waves, for a total of 2,700 interviews.

To supplement the Use and Usage Survey, the ORETF provided questions for the National Gardening Association survey . The survey was conducted at the end of the 1998 calendar year as a face-to-face, in-home personal interview with 1,500 households nationwide. The information collected provided additional data on garden and lawn sizes, the time of year and total time spent in gardening activities, and the pesticide use on trees.

Conclusion

The ORETF established a standard method to measure turf transferable residue that is now a required methodology for regulatory guideline studies. The data generated by the ORETF was incorporated into the 2012 USEPA Standard Operating Procedures for Residential Pesticide Exposure Assessment (*1*). The ORETF achieved its objectives and developed data that met requirements by working collaboratively as an industry and with regulatory authorities through the JRC. The formation of the ORETF allowed for member companies to share scientific expertise and financial resources to develop a more consistent and comprehensive exposure database than could be developed by any individual member company. The formation of a task force also allowed member companies to work as a single entity with regulatory experts to improve efficiencies and ensure the data developed would meet regulatory needs.

The work of the ORETF highlights the major benefits to a generic approach, including greater confidence and improved quality and consistency in exposure estimates; enhanced ability to identify and address critical gaps in exposure data; greater ability to inform regulatory policies and decision making with respect to handler and reentry safety; and a significant reduction in data development costs and improvement in the quality and predictability of product registrations.

Acknowledgments

The author would like to thank the 33 Member Companies with special acknowledgment for the individual member company representatives for their support, efforts, and participation to develop and conduct the ORETF program. Thank you to the JRC members who guided the ORETF to ensure sound science was applied and quality data were generated. A very special thanks to the ORETF Task Force Manager, Dr. Dave Johnson, and Task Force administrative assistant, Rhonda Bischel, for all of their assistance and review. Additional thanks to Dr. Monty Eberhart for his guidance and review of this chapter.

References

1. US Environmental Protection Agency 2012 Standard Operation Procedures for Residential Pesticide Exposure Assessment. https://www.epa.gov/pesticide-science-and-assessing-pesticide-risks/standard-operating-procedure-residential-exposure (accessed 2021-03-01).
2. Hackathorn, D.; Eberhart, D. Data-Base Proposal for Use in Predicting Mixer-Loader-Applicator Exposure. In *Dermal Exposure Related to Pesticide Use: Discussion of Risk Assessment*; Honeycutt, R. C., Zweig, G., Ragsdale, N. N., Eds.; American Chemical Society Symposium Series 273; American Chemical Society: Washington, DC, 1985; pp 341–355.
3. Klonne, D. Summary of Exposure Data for all Mixer/Loader/Applicator Studies Conducted or Purchased by the ORETF. 2002. ORETF Report Submitted to USPEA as MRID 45663701.

4. Klonne, D.; Cowell, J.; Mueth, M.; Eberhart, D.; Rosenheck, L.; Ross, J.; Worgan, J. Comparative Study of Five Transferable Turf Residue Methods. *Bulletin of Environmental Contamination and Toxicology* **2001**, *67*, 771–779.
5. Rosenheck, L.; Cowell, J.; Mueth, M.; Eberhart, D.; Klonne, D.; Norman, C.; Ross, J. Determination of a Standardized Sampling Technique for Pesticide Transferable Turf Residues. *Bulletin of Environmental Contamination and Toxicology* **2001**, *67*, 780–786.
6. Fuller, R.; Klonne, D.; Rosenheck, L.; Eberhart, D.; Worgan, J.; Ross, J. Modified California Roller for Measuring Transferable Residues on Treated Turfgrass. *Bulletin of Environmental Contamination and Toxicology* **2001**, *67*, 787–794.
7. Ross, J.; Driver, J. Refining Estimates of Potential Hand-to-Mouth-Based Incidental Ingestion Exposure: Frequency of Hand to Mouth Contact. 2003. ORETF Report Submitted to USEPA as MRID 46042401.
8. Ross, J.; Driver, J. Refining Hand to Mouth Exposure Estimates: Finger Surface Area Contacting Mouth. 2003. ORETF Report Submitted to USEPA as MRID 46042402.
9. Klepeis, N.; Nelson, W. C.; Ott, W. R.; Robinson, J. P.; Tsang, A. M.; Switzer, P.; Behar, J. V.; Hern, S. C.; Engelmann, W. H. The National Human Activity Pattern Survey (NHAPS): A Resource for Assessing Exposure to Environmental Pollutants. *Journal of Exposure Analysis and Environmental Epidemiology* **2001**, *11*, 231–252.
10. Baugher, D.; Klonne, D.; Mihlan, G.; Ross, J. The ORETF Algorithm for Defining the Relationship of Transferable Turf Residues to Post-Application Dermal Exposure. 2004. ORETF Report Submitted to USEPA as MRID 46190501.
11. Johnson, D.; Thompson, R.; Butterfield, B. Outdoor Residential Pesticide Use and Usage and National Gardening Association Survey. 1999. ORETF Report Submitted to USEPA as MRID 44972202.

Chapter 7

The FIFRA Endangered Species Task Force: Dealing with Unusual Challenges and Multiple Agencies

Bernalyn D. McGaughey*

FIFRA Endangered Species Task Force / Compliance Services International, 7501 Bridgeport Way West, Lakewood, Washington 98499, United States

***Email: bmcgaughey@complianceservices.com**

The FIFRA Endangered Species Task Force (FESTF, with the embedded acronym FIFRA referencing the Federal Insecticide, Fungicide and Rodenticide Act) was formed in December of 1994 to develop data required by the Environmental Protection Agency (EPA) of applicants or registrants relative to the potential impacts of pesticides on species listed under the Endangered Species Act (ESA) and having a status of threatened or endangered (listed species). FESTF was organized by pesticide registrants to develop information sufficient to address FIFRA data requirements imposed on these registrants. The work accomplished by, and continuing through, the Task Force has provided necessary information for many pesticide product registrations and re-registrations in the United States, but via a path not initially anticipated under the typical expectations for Task Force data development. The Task Force covers a wide cross-section of pesticide registrants that benefit from this endeavor and has submitted multiple datasets to EPA and other federal Services (e.g., the Fish and Wildlife Service – United States Department of the Interior [USDI], National Marine Fisheries Service – National Oceanic and Atmospheric Administration [NOAA]). FESTF has also developed a customized information management system that allows access to and comparison of data in a spatial and tabular software environment. The original mission of the was envisioned to be addressed through a Cooperative Research and Development Agreement (CRADA) for the purposes of EPA's state by state assembly of data for listed species. However, the CRADA proved unworkable, and rather than EPA being burdened by the task of data assembly, FESTF took on the effort and commitment to support data aggregation and software program development. This chapter will explore the challenges that required unique approaches to data provisioning and will assess how 25 years of operations, constantly changing data, shifting policy, and advancing technology have been managed to address the complex data interactions brought about by FIFRA Registration Review and subsequent ESA consultation, and ultimately expected to be adopted routinely into the overall FIFRA registration process.

Some Background on the FIFRA/ESA Process

If ever there were two regulatory acts developed in absence of a concept of how they would interact with one another, it is the Act that controls pesticide use and regulation, the Federal Insecticide, Fungicide and Rodenticide Act (FIFRA), and the Act that protects species at critical risk, the Endangered Species Act (ESA). In what is now a rather elegant process, FIFRA works from the national level to register a pesticide whose use is governed by a label wherein "the label is the law." To carry out the implementation and enforcement of the label, a network of federal and state organizations fills roles that inform, recommend, monitor, and enforce pesticide use. The aforementioned "elegance" of the FIFRA regulatory system is that it allows a national federal registration action to trigger implementation and specific use patterns that employ other federal agencies, such as the US Department of Agriculture (USDA), universities and states in tailoring and enforcing pesticide use in a way that is safe and responsive to local needs. Conversely, in a process that is designed for and centered on the local scene, and carried out largely independently by the organizations responsible for administering the Act (the US Fish and Wildlife Service, under the Department of Interior, and the National Marine Fisheries Service, under the Department of Commerce, collectively the Services), actions under ESA are typically "between locals" – local Services experts, local agencies involved in an action, and local entities proposing to perform the action. In the FIFRA/ESA intersection, when the action of registering a pesticide requires the action agency (in this case, the Office of Pesticide Programs in the Environmental Protection Agency, EPA) to consult about possible species harm, wherein the Services must determine the potential extent of that harm, finding common operational ground is difficult. Since the passage of the ESA in 1973 and the Tellico Dam decision (*1*), until now (See Figures 1 through 4) and beyond (see Figure 5), affected parties continue work to establish how pesticide consultation should function. The Tellico Dam case was a United States Supreme Court case and the Court's first interpretation of the Endangered Species Act of 1973. It set precedent for ESA prevailing, through consultation, over any federal action that might impact listed species. With this decision, pesticide registration consultations began, because the registration of a pesticide label is considered a "federal action."

It was within this quandary of emerging methodologies from which the FIFRA Endangered Species Task Force arose.

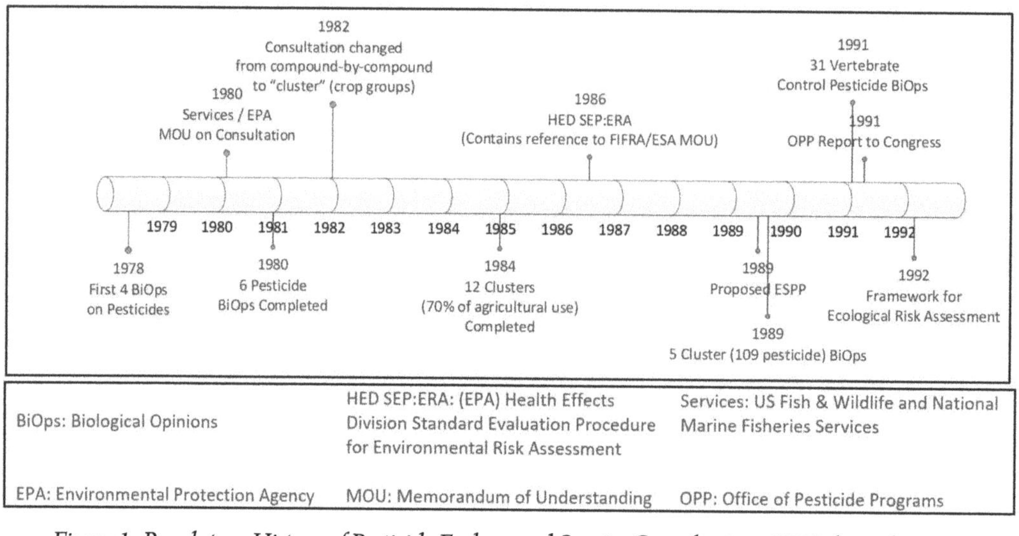

Figure 1. Regulatory History of Pesticide Endangered Species Consultation, 1978 through 1992.

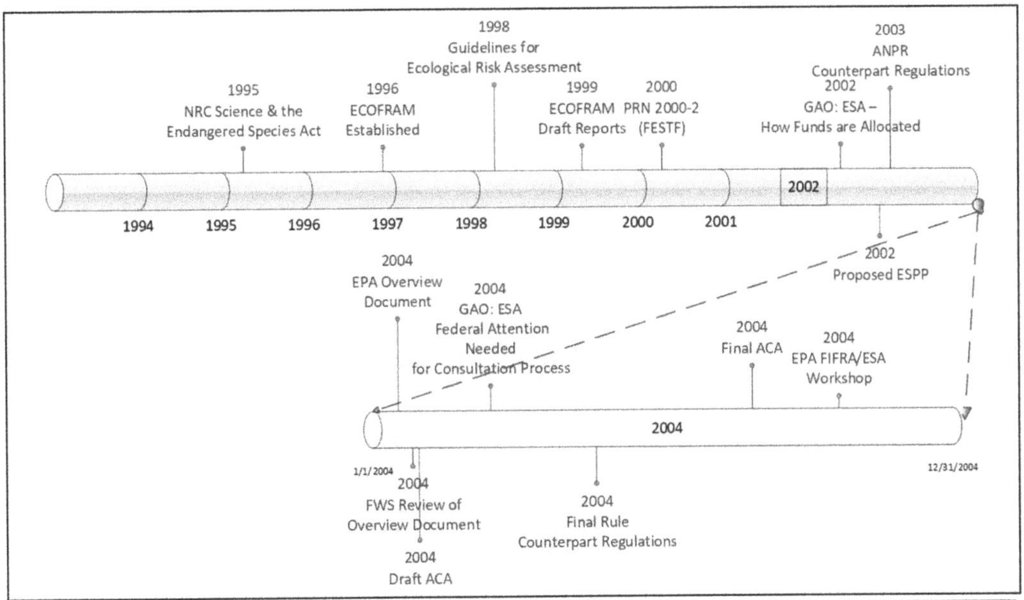

EPA: Environmental Protection Agency	GAO: Government Accountability Office	FWS: US Fish and Wildlife Service
ECOFRAM: Ecological Risk Assessment Framework	ANPR: Advanced Notice of Proposed Rulemaking	ACA: Alternative Consultation Agreement

Figure 2. Regulatory History of Endangered Species Consultation, 1993 through 2004.

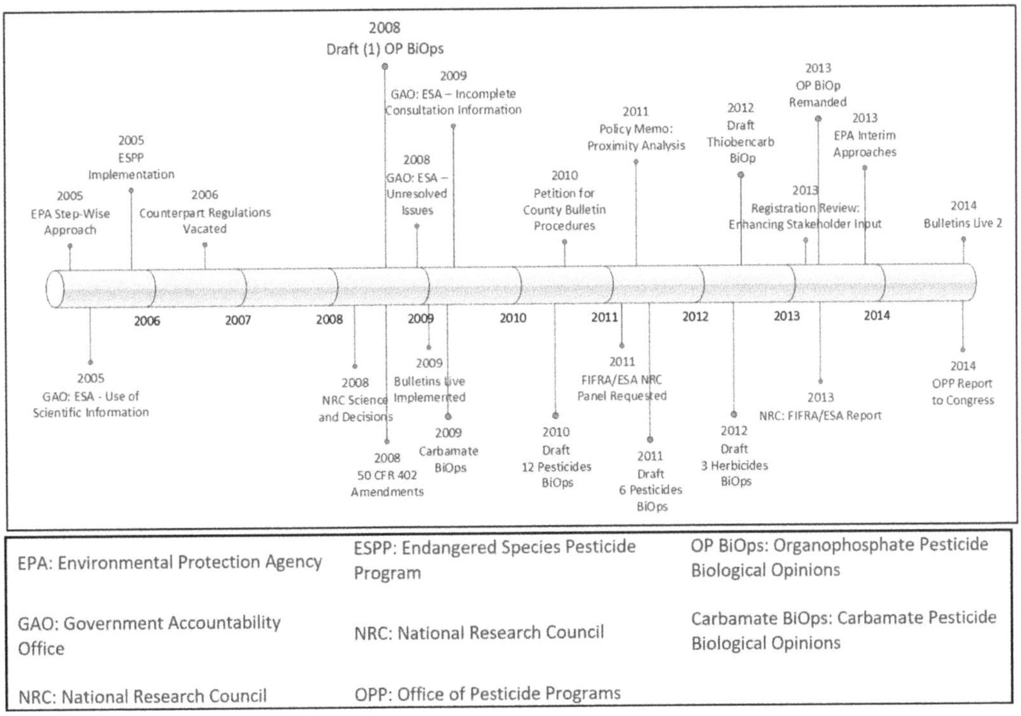

EPA: Environmental Protection Agency	ESPP: Endangered Species Pesticide Program	OP BiOps: Organophosphate Pesticide Biological Opinions
GAO: Government Accountability Office	NRC: National Research Council	Carbamate BiOps: Carbamate Pesticide Biological Opinions
NRC: National Research Council	OPP: Office of Pesticide Programs	

Figure 3. Regulatory History of Endangered Species Consultation, 2005 to 2014.

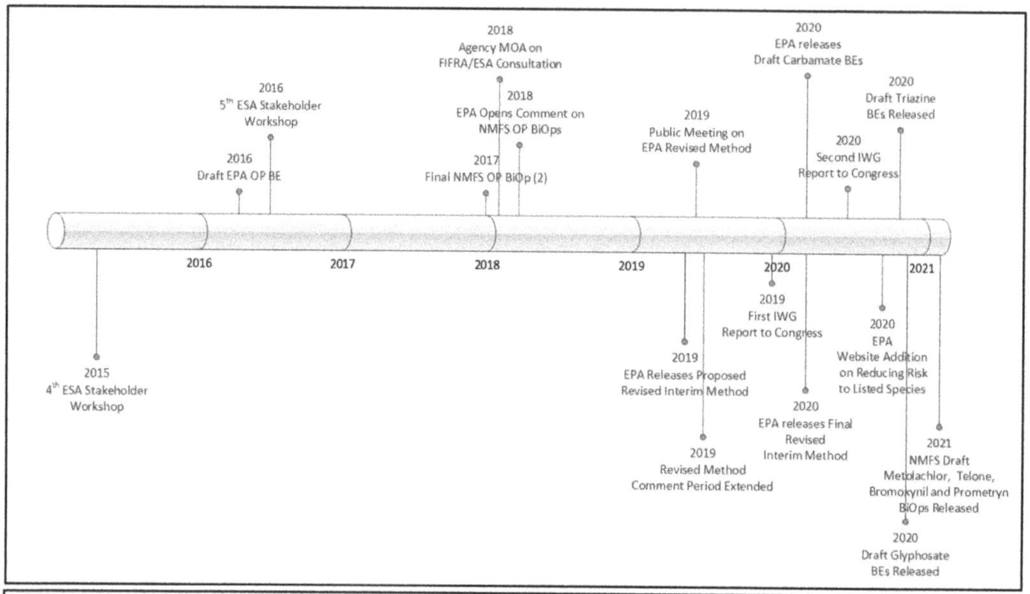

See Figures 2 and 3 for acronyms

Figure 4. Regulatory History of Endangered Species Consultation, 2015 through February 2021.

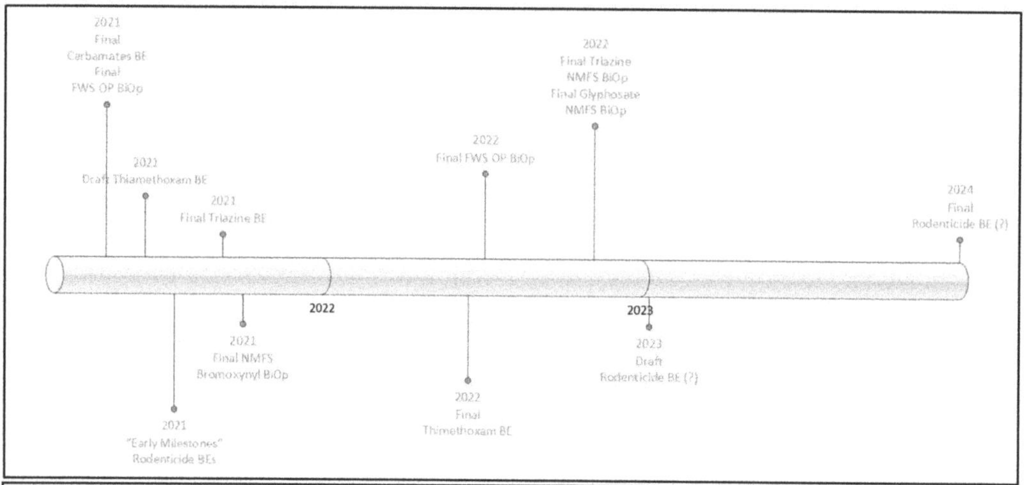

See Figures 2 and 3 for acronyms. Note the two FWS BiOps; the first was not fully released in 2021 and instead rescheduled for 2022

Figure 5. Currently Scheduled Consultation Actions, March 2021 through 2023.

From the many iterations of risk assessment method development and consultation approaches, litigation over the process of consultation for pesticides has proliferated. The FIFRA/ESA interface is particularly vulnerable to process-oriented litigation because of the mismatch (local "projects" under ESA versus the national action of registering a pesticide) in the operating environment of the two laws. Additionally, pesticide science is extremely complex and multifaceted, but poorly understood by those not experienced in it, which means it is difficult to communicate in a court or public setting.

The FESTF has not been engaged in addressing public opinion, or on the development of or commentary on policy; the mission of FESTF is meeting EPA data requirements through data aggregation and provision so that the best available data can be used by federal agencies to support their pesticide endangered species risk assessments regardless of the evaluation methods that emerge under policy. But, understanding the complexity of the challenge of assessment, and the approaches and difficulties faced by the agencies over the years, highlights the importance of being able to deliver high-quality, well-documented, defensible, transparent, and repeatable endangered species effects determinations and risk assessments.

Table 10. Results of the T&E species and AL-I land use proximity analysis for

The highlighted EOs were eliminated by Tier A. - denotes "no exact data"; NA denotes "Not Available"; and ~ denotes "area covered below"; MOC denotes "mapped out of county", PWP denotes "point within or near a polygon". Point denotes "EO mapped as point", p,xyzac denotes "EO mapped as polygon of xyz acres".

	County	Species	Mapped As	Certainty Category	EO Code	Agricultural Acres in Proximity Area	
						300 Ft	1,320 Ft
1			Point	HIGH	102	~	~
2			p,13ac	HIGH	103	~	~
3			p,8ac	HIGH	104	~	~
4			p,31ac	HIGH	105	62.3	224.7
5			NA		EPA	-	-
6			-		42	-	-
7			MOC	LOW	19	~	~
8			p,10ac	LOW	45	4.2	88.3
9			-		2	-	-
10			NA		EPA	-	-
11			NA		EPA	-	-
12			NA		EPA	-	-
13			NA		EPA	-	-
14			NA		EPA	-	-
15			NA		EPA	-	-
16			NA		EPA	-	-
17			Point	HIGH	7	3.4	93.8
18			NA		EPA	-	-
19			NA		EPA	-	-
20			NA		EPA	-	-
21			NA		EPA	-	-
22			MOC	HIGH	106	0.0	6.4
23			NA		EPA	-	-
24			-		2	-	-
25			-		38	-	-
26			-		39	-	-
27			NA		EPA	-	-
28			NA		EPA	-	-
29			NA		EPA	-	-
30			~		100	~	~
31			p,28ac	HIGH	101	8.4	32.2
32			Point		15	-	-
33			NA		EPA	-	-
34			p,103ac	HIGH	10	~	~
35			p,8ac	HIGH	18	~	~
36			p,2ac	HIGH	28	~	~
37			p,8ac	HIGH	44	0.0	0.3
38			-		43	-	-
39			NA		EPA	-	-
40			NA		EPA	-	-
41			NA		EPA	-	-
42			Point	LOW	1	0.0	8.2
43			NA		EPA	-	-

Figure 6. Example table from FESTF 1996 feasibility study (redacted to remove confidential information).

What is the FIFRA Endangered Species Task Force?

FESTF is a data development organized to address endangered species location data requirements for pesticide product registrations. Initially, those data requirements were stated by EPA as providing proximity data that would allow EPA to determine which, if any, listed species were near a potential labeled use site. Once EPA knew the species of concern, then they could apply their usual risk and label mitigation process to protect those species from potential harm from labeled pesticide use. In addition to providing such data, the goal was to develop a means to manage the data over time, and to quality test those data to prove their reliability.

As a consortium responding to data requirements, FESTF, as an independent limited liability corporation, could serve as a scientific organization to gather and manage relevant data under the supervision and management of its membership. In this way, the needs of EPA could be met by the Task Force's provision of work products and any pesticide product registrant could either participate as a FESTF member in the original development of the data to fulfill data requirements supporting its registrations, seek membership in the at a later date and thereby secure rights to rely on those previously generated data, or cite FESTF data as a third-party using the terms provided by FIFRA data compensation provisions. Like other industry s, FESTF allowed its members to develop a single agreed-upon and uniform response to EPA's data requirements, while acting as a single entity to communicate with EPA in fulfilling those data requirements and avoiding unnecessary redundancies or duplicate efforts.

Robust industry efforts to examine EPA's endangered species data needs began in 1993 when the first requirements for species proximity data were issued by EPA as conditions of registration to several registrants. Initially, these conditions varied from registration to registration and opened the possibility that data requirements would not be equitably assessed among registrants. The conditions also presented endangered species data requirements as a conceptual matter, without outlining any consistent or predictable means by which to meet them: Proximity of listed species to what? Was this a modeling exercise or are real-world relationships needed? How should data be presented and what make the data useful is risk assessment? Were good quality data even available?

Proximity of Listed Species to What?

Before the FESTF was formally incorporated, CropLife America (then known as the American Crop Protection Association) formed a work group to develop a demonstration project that would explore a means of mapping listed species and their locations in relation to agricultural crops. A paper-based feasibility study was prepared for the Midwest on a limited number of species and crops and submitted to EPA in 1996. But what was the agency to do with a cabinet full of annotated cartographic depictions of the known or estimated location of a species? And since crops rotate, species listings change and grower planting choices vary, what aspect of proximity is defined by a collection of static, point-in-time maps? If nothing else, the feasibility study demonstrated that relationship analyses of species to pesticide use sites, and risk conclusions based on those analyses, were not as simple as it might seem. The most typical way to present results given the technologies of the time, was in tables. Figure 6 is a scanned copy of a table in the final summary report for the feasibility study (2), with confidential information redacted. It is included here for two reasons: one, as a reminder of the rigidity of the day, over 25 years ago, that demanded results be tabular in most cases when the outcome of mapped or spatial comparisons were reported; and, two, to demonstrate that interpretation of the information and its application to evaluating species exposure or risk was complex (see, in Figure 6, the columns marked "mapped as" and "certainty category").

The complexity of these assessments continues to present a challenge to the FIFRA/ESA consultation process today.

Was This a Modeling Exercise or Are Real-World Relationships Needed?

Other complex relationships that require consideration in pesticide risk assessment have, over time, been subject to the development of predictive models: drift, runoff, dietary exposure, and other environmental models are a practical, if sometimes assumptive, way to estimate environmental exposure. However, the models that have been developed can rely on the entry of parameters such as windspeed, dissipation, residue data and values for the environmental variability that influence them. There are several reasons why the development of a model to predict a given listed species exposure is neither predictive nor, from a regulatory perspective, meaningful. Some of these are:

- Complexity: There are over 1600 species and their character crosses nearly all taxonomic categories, from aquatic invertebrates to large mammals. Often their habitat requirements are very specific, obligatory, or unique.
- Supporting Data: Even though the Services have stewarded listed species for more than 40 years, the knowledgebase on each species runs from well-studied to unknown. Some listed species are of such scarcity or obscurity that very little at all is known about them.
- Constantly changing data: As crop data is collected – today via remote sensing and other means – and species listings are decided, the elements important to the relationship of a pesticide use to exposure of a listed species also change. A static map or dataset has a limited useful lifetime.
- The nature of the Endangered Species Act: Unlike the way FIFRA interpretation and evaluations have evolved with time and scientific advancements, ESA assessments are assigned a more restrictive regulatory definition. Essentially, two questions must be answered, (1) "Will even one individual in this species be harmed, or its habitat adversely affected, if this product is used in this field on this day," and (2) "If this species is likely to be harmed, will its loss jeopardize the species as a whole?" If one is building a parking lot, and can survey the ground and surrounding area, these questions are relatively easy to answer, although controversies over the interpretation of harm do still arise. These questions, when transferred to a national decision, are overwhelmingly complex – as the long struggle for successful pesticide consultation demonstrates.

How Then Should Data Be Presented and What Make the Data Useful in Risk Assessment?

The 1996 feasibility study demonstrated that a filing cabinet full of annotated maps could not deliver to the agencies the information they needed. Once EPA had the opportunity to examine the results of the feasibility study, the agency and industry recognized that more time was needed to identify the steps of a listed species evaluation and what it took to inform those steps. There was not a formulated way to apply "proximity" to species risk. To address the need for data to inform the assessment, EPA and FESTF met in a series of workshops to better understand just what data were required, what the steps were in EPA's species evaluation process and which EPA staff were involved in them, and in what form were results meaningfully presented.

Were Good Quality Data Even Available?

As the workshops delved into these questions, it quickly became obvious, as the available data were examined and information needs addressed, that constantly changing data and the complexity of the reviews under FIFRA combined with consultation under the Endangered Species Act required more than a one-time aggregation of location and crop data. It also required multiple levels of review, starting with the "usual" FIFRA risk assessment process but moving to species review by multiple levels of staff, at that time in EPA the Field External Affairs Division (FEAD) as well as in the risk management divisions (Registration Division, Pesticide Re-evaluation Division, and others), and in the Services, consultation biologists at both National Marine Fisheries Service (NMFS) and the Fish and Wildlife Service (FWS). As various work groups walked through each aspect of need, the three basic work products of FESTF began to take shape: reliable data sources, a system to manage them and a means to demonstrate their quality and reliability. Interestingly, there was no reliable, let alone aggregated, source for the types of data and information required for FIFRA/ESA species assessment.

Why Was FESTF Formed?

In short, FESTF was formed to develop or collect the data necessary for closing the gap between the national FIFRA risk/benefit approach and the ESA mandate to ensure "no harm" to listed species. Rather than individual registrants submitting various data collected or generated under equally varying means, the Task Force could focus on the "data need" for the assessment process rather than the "registration need" of an individual product. Then, having assembled the best available data, the Task Force can ensure those data could be applied consistently to evaluations, updated uniformly, and accessed from a common platform.

An industry Task Force provided the structure to explore, discuss, formulate, and build a uniform collection of data. As how both EPA and the Services processed their regulatory obligations related to listed species became better understood through agency meetings, discussions with FIFRA and ESA scientists, and research on existing data sources, the Task Force could also unify response to those needs by delivering the aggregated data and providing a system that could hand off one level of assessment to the next group of agency staff needing to build upon it.

The aggregation of data on species locations was particularly challenging but understanding those data before designing a system to manage them was critical to success. Data depicting species locations were not at that time available from a central source at the Services nor were (or are) they available for each species. Consequently, EPA envisioned obtaining access to data of this type from a new data distribution platform planned by The Nature Conservancy (TNC). Access to that data was specified in FESTF's data requirements as embodied in Pesticide Registration Notice 2000-2 (3). The TNC data had been gathered over time as their conservation and land acquisition activities were carried out and grew to have usefulness on its own merit. At first, the distributing arm of these data was not described. Most of the data were held by Natural Heritage programs, a network of state organizations originally funded by TNC but eventually either independent non-profit organizations or absorbed by and incorporated into state agencies. An umbrella for these separate entities that would allow them to uniformly distribute data on a national platform did not exist.

One possible way of developing that platform, and EPA's access to it, was through a Cooperative Research and Development Agreement (CRADA), which allows federal and non-federal parties to join in conducting research. A CRADA is defined as "Any agreement between one or more federal facilities and one or more non-federal parties under which the government, through its facilities, provides personnel, services, facilities, equipment, intellectual property, or other resources with or without reimbursement (but not funds to non-federal parties) and the non-federal parties provide funds, personnel, services, facilities, equipment, intellectual property, or other resources toward the conduct of specified research or development efforts which are consistent with the mission of the facility; except that such terms do not include a procurement contract or cooperative agreement

(4)." Under a CRADA, the multiple government agencies involved in FIFRA/ESA consultation and industry as represented by the FESTF, could combine research efforts to establish a distribution (information management) system that would deliver state data formatted uniformly for a national depiction of species locations.

This pushed the edge of the envelope from both an interagency process standpoint as well as an information technology standpoint. Geographic information systems (GIS) were certainly operational at that time (the mid- to late-1990s), but not unified and still requiring computing power beyond that typically found now in our own personal computers. This and other complexities prevented the execution of a CRADA, despite a strong collective effort over four years to form one. When it became clear that a CRADA was no longer the instrument of the agency's choice, FESTF took on the task of accessing and aggregating the TNC data directly so that the aggregated data could in turn then be accessible to members for submission to EPA as required under FIFRA to support the members' registrations. This task and FESTF's ability to commit to it was another demonstration of the benefit of s: one scientific entity could concentrate on the data provision need and work out differences in ideas and approaches presented by members and combined for the best possible approach with respect to data quality and reliability, consistent data access and efficiency of resources.

During the time of exploration of a CRADA, TNC itself was evolving its concept of data delivery and access through various platforms. While various avenues were explored, the final venue was a new data-providing organization, NatureServe. FESTF met not only with the federal agencies about meeting the data requirements for listed species evaluation, but also met with TNC and then NatureServe to understand what was or would be available to assemble as a portion of the aggregated national dataset.

Many Minds, One Guiding Concept

Through this research and exploration, the stable "three-legged stool" emerged: a compiled national dataset gathered exclusively for FIFRA/ESA assessments; a management system to house them that ensured quality and reliability of data; and output for the agencies that met their own, still-evolving process. Returning to the management side of the information, FESTF conceptually constructed the operations and data necessary for a data management system. While much of the information needed was spatial, it was decided, largely due to the technical and cost limitations attached to it, that a spatial, GIS-based system was not yet the tool that EPA needed. Instead, the FESTF information management system was built around the procedural and technical method that EPA used to address listed species in the review process for registration. This meant capturing the (1) outcome of the Environmental Fate and Effects Division's (EFED's) risk assessment, (2) resulting EFED-proposed mitigations, (3) real-world adjustments by species experts in FEAD, (4) state or registrant-submitted species data and assessments, (5) informal oversight by Services, which at the time were operating under an interdepartmental memorandum of agreement (5), and (6) final acceptance and approved labeling by the Registration or Reregistration Divisions (PRD or RD), and storing of both unaccepted and accepted mitigation statements. Over the ensuing 2 to 3 years the information management system (IMS) was developed, tested, and finally applied to three test cases. Following that, two assessments on compounds under review at the time (2,4-D and atrazine) were run through the process at a screening level of effort – meant to elucidate the process – and submitted to EPA. EPA provided data evaluation reviews on those two assessments (6), and the data reviews helped FESTF address additional needs or concerns of the agencies. From the point of establishing the concept for meeting data requirements to the actual review of preliminary work products, about

8 years had elapsed – but the regulatory process moves slowly and from a development process, the actions described to this point were still "immature."

The EPA methods for species assessment were, and still are, evolving – presenting another challenge to FESTF with respect to delivering data in a way that met the approaches decided upon by EPA scientists, or as demanded by policies set for the EPA scientists to follow. As time moved on after the initial submission and activation of the IMS "version 1," layers of its capabilities became obsolete or not relevant to a new approach dictated by EPA methodology or policy. FESTF was not always advised of these changes in advance, but adapted data aggregation and delivery through new versions of the IMS when changes were required to keep the system useful and effective in meeting data needs. Such evolution is universal in information technology since computer science advances rapidly, data grow and change over time, and government policy – more slowly, but just as significantly – evolves over time. And with change upon change, eventually a revamp is needed.

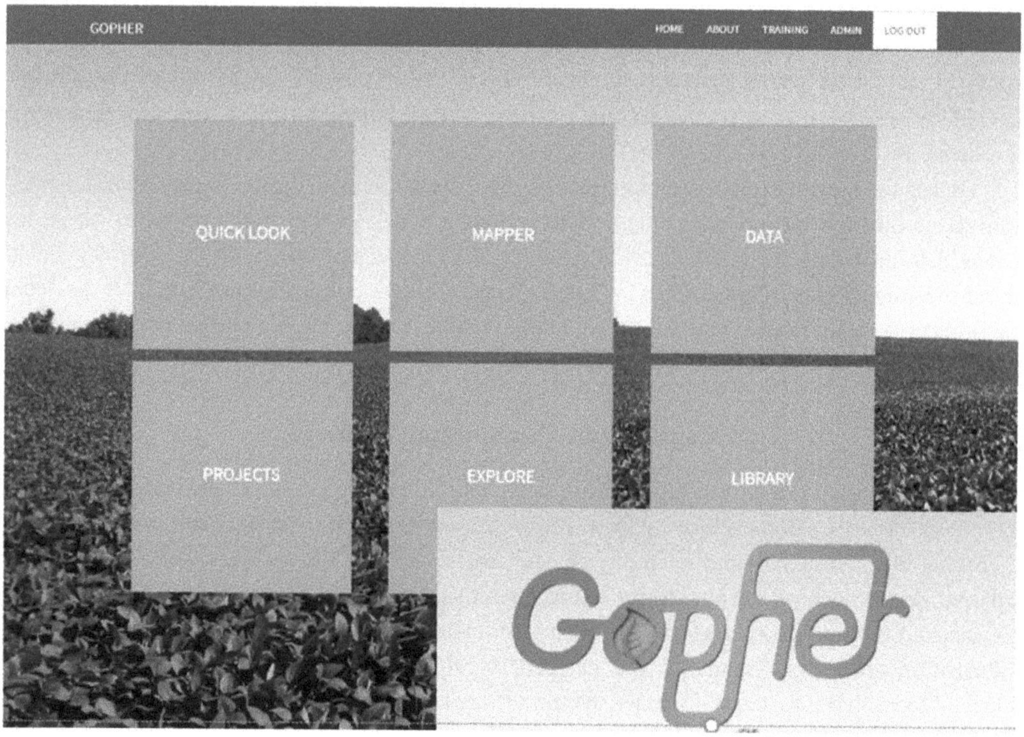

Figure 7. FESTF's "Gopher" logo and home page.

Approximately 10 years into the life of the Task Force, changes were so significant that the IMS was entirely re-thought and design was changed to a spatially based operating system still capable of storing and retrieving tabular data, housing custom proprietary datasets of members and making spatial comparisons. At long last, the IMS deserved a name: Gopher (Figure 7). It was even given a logo, which reflects the ability of the system to allow users to tunnel different routes through the data to produce work products on-demand by querying data in various structured or free-form ways. The rigid structure of the first IMS was abandoned, and Gopher was designed to be flexible and easily modified for changes in assessment policy or data sources. Gopher has now been made available to EPA, FWS, NMFS and USDA as well as members of FESTF. Output from Gopher has been used in agency submissions, scientific conferences, agency meetings and to address certain questions from stakeholders such as Defenders of Wildlife, CropLife America, pest control associations, and state agencies. The collective Task Force approach allows varying levels of needs, not only between different agencies but also between members, to be met equitably. The six tiles of entry into Gopher shown in Figure 7 (Quick Look, Mapper, Data, Projects, Explore, Library) reflect these levels:

- **Quick Look** is designed for users not having extensive spatial data manipulation skills and gives a snapshot of species and pesticide use site relationships.
- **Mapper** allows users with limited knowledge of spatial data to select features to combine and map and subsequently apply to answering questions about features and layers, such as whether a species is located upstream and away from pesticide use sites, as shown by the example in Figure 8.
- **Data** is the access point to the of 20+ datasets housed in Gopher, each of which can be downloaded and/or further explored, combined, and "mined" for information to inform the conditions and facts related to listed species and pesticide use sites. This is also the section of the system allowing upload of new datasets, which can be flagged for use by all or kept proprietary to a given Gopher user or company.
- **Project** or the "**Project Toolkit**" is the primary container for work done within the GOPHER IMS. A project contains important information regarding the assessment being completed such as the product being tested, a description of the application, the active ingredient associated with the product, and the list of users who have access to the project. Projects also contain place to include a narrative description of the work being done on the project. Project files may be stored here for handy reference of materials used in or produced by a given user's project.
- **Explore** allows the data user to construct queries of various data sources, such as the FWS-based Status of the Species reports, NatureServe's multijurisdictional dataset, topics such as habitats, species habits and pesticide use sites, or user-developed projects proprietary to a given company or user.
- **Library** allows the user to search for documents or contact information for experts, all completely searchable and housed as links in Gopher. Documents can be retrieved by file name, author, citation, etc. and contacts can be searched by first or last name, occupation, or organization. New documents and contacts can be uploaded to Gopher and become part of the reference resources available to the user.

All data uploaded to Gopher are quality checked for completeness and accuracy *with respect to their reflection of the source data.* Even though during data aggregation and use, FESTF members and staff find errors in the data the data housed in Gopher are not changed from those reflecting the parameters of the source data. Known errors are noted in metadata, but the databases themselves are not altered in any way. Thus, the aggregation of data, steps to make them compatible with the family of data housed in Gopher, quality checking and the Gopher system itself become the work products of FESTF.

Who Are FESTF Members?

FESTF is open to all pesticide registrants. Some FESTF member companies joined the Task Force at the time of its original formation, while others became members later in accordance with the FESTF LLC Agreement, subsequently purchased a membership from a company leaving the business or acquired a membership through corporate mergers and acquisitions. Membership has changed over time for these reasons. Members as of February 2021 are listed in Figure 9. Additionally, hundreds of companies have made offers to pay compensation for reliance on FESTF data for more than 1000 products, consistent with FIFRA's data protection and citation requirements.

Figure 8. A map generated using Gopher, showing that the Austin blind salamander is located well away from cultivated croplands.

ADAMA Agricultural Solutions	Nichino America
AMVAC Chemical Corporation	Nippon Soda Company
BASF Corporation	Nissan Chemical Corporation
Bayer US Crop Science	Nufarm Americas, Inc.
Corteva Agriscience	PBI Gordon Corporation
Corteva Dow Operations	Syngenta Crop Protection
FMC Agricultural Products	Valent USA LLC
Gowan Company LLC	UPL Ltd.
ISK Biosciences Corporation	

Figure 9. FESTF member companies as of February 2021.

What Data Have FESTF Generated?

FESTF is a "data generation Task Force" whose purpose is to meet federal regulatory requirements to support its pesticide registrations, and its mission has been to gather and aggregate data rather than to conduct original research that generates new data, as some s do. FESTF also provides a means to store, maintain and deliver information in or from the housed data. As such, FESTF's data resources increase as they are found to be relevant and acceptable for use in species assessment processes. Data on listed species is distributed across both national and local platforms. Some species are well studied and documented while others are poorly understood. Consequently, available data on each listed species can be highly variable. As mentioned previously, some species have no range data while others are well described or mapped. Many details on species locations and populations are considered sensitive so they are not available to the public or for assessment purposes. Data on crops can come from multiple sources – and may or may not corroborate from one source to another. It therefore takes knowledge of both what data are available to the reviewer as well as what the limitations of the datasets are to apply them properly in an analysis. FESTF's Gopher provides the capability of comparing these datasets, but the user must understand what they are comparing and what conclusions can be drawn from any comparisons they make. Thus, Gopher provides the freedom to select the dataset(s) of interest for any one assessment process, or to add new datasets as they are found to be robust and useful.

In FIFRA/ESA endangered species assessment, data necessary to compare species locations and potential exposure are often not easily joined. Nomenclature, update intervals, spatial portrayal and other variables all need to be reconciled so that efficient and representative comparisons can be made. These reconciliations are made when data are uploaded in Gopher. Gathering, delivering, and keeping the data useful and current is iterative and changes over time, thus these efforts also become a "work product" of FESTF.

What Is Unusual about FESTF?

Many task forces have had to deal with more than one agency, but in most if not all other cases, these were agencies having a prescribed role in pesticide regulation, such as the Food and Drug Administration's (FDA's) responsibility to evaluate and manage contaminants in food, including the management of pesticide residues in food and feed items. However, in this example of FDA/EPA interaction, EPA is delegated the authority to act for FDA by the Federal Food, Drug and Cosmetic Act (FFDCA), rather than an EPA assessment being handed off to FDA for further analysis and deliberation. In the case of endangered species, two acts never connected by common operating language or definitions separately direct risk evaluation, with one set of end points and definitions under FIFRA and a second set under ESA. To complicate things further, depending on the species and its defined position as a protected organism, its protection under the ESA is managed by NMFS (generally wide-ranging species associated with marine life stages, or life stages in or associated with navigable freshwaters) or FWS. These two agencies have different approaches and resources with which to evaluate species risk from the action of registering a pesticide, and each agency's needs and operating environment must be addressed when data are provided.

Also, somewhat unusual in an established Task Force is the lack of a specific data target or problem to solve. Challenges like development of a spray drift model and data to support it, or evaluating residential exposure using methods based on exploratory studies are indeed large and complex tasks, but at least their development and iteration fit under the "FIFRA way of doing risk assessment." That "FIFRA way" at least provides guideposts for agency expectations and Task Force outcomes. When FESTF was formed, there was not an existing basis for providing data and information needed in the FIFRA/ESA consultation process, and today even yet the process is "interim" and repeatedly challenged legally, forcing further changes in the methods applied. The marriage of two unique environmental laws is yet to happen fully and successfully, but the concept of data aggregation and management as first laid out by FESTF members remains viable and relevant. Throughout the evolving process and advancing technology, FESTF has built on its initial efforts, enhanced them, and adjusted them in response to policy, data, and technological change.

FESTF did not develop a model or complete a study-based research effort. The heart of FESTF's effort is the aggregation of data, keeping it current, and delivering it reliably and in a way that maximizes efficiency in data use, application, and interpretation.

Data are constantly changing. Species are listed and delisted, or the description of their range is modified. Their populations change or population counts are adjusted over time and their recovery potential, until delisted, is never static. Thus, stressors to them are evaluated in different comparative settings from one time to the next. Data sets are improved, corrected, or added to over time. Cropping trends change, temporarily or according to definite trends over time. Information technology increases the ability to deal with larger data sets, higher refinement, and faster processing, but does not necessarily inform or impact assessment policy. Instead, this variability is addressed by careful reliance on and access to current and relevant, best available data, coupled with the best information management technologies to efficiently manage and deliver them.

What Challenges Has FESTF Faced?

The data challenges mentioned in the previous paragraph have driven the design and work of FESTF. The challenges faced in doing this work include the need to serve a process in absence of established policy or protocol. Thus, to the best ability of the collective member-minds that have built FESTF resources, species assessment data and process needs must be anticipated so that data and systems serve them as agency methods pass through multiple iterations responsive to trying to make FIFRA and ESA work together. This dynamic scenario demands that concepts envisioned must be turned into reality in an iterative, stepwise fashion. Even then, a change in administrations or a court decision setting precedent can be game-changing. Staff and managers change, collective knowledge is lost or must be re-explained. Operating effectively in the FIFRA/ESA nexus requires understanding the culture in and regulatory drivers for each agency involved in the process. It also means preserving the ability of the pesticide user to manage lands in their care for both pest protection and listed species stewardship.

Centralizing decentralized data, where there is no "final product," is a second major challenge. In overcoming this, data must be accepted or proven as best available. Data must also be knitted together even if common data elements have different names – even what may seem as simple as a species name can vary from one data set to another. Spatial data must be projected commonly so it is comparable from one dataset to another, and the power of its resolution must be accounted for and limitations understood. Operating in a Task Force environment actually makes this as "easy" as possible because the differences, nuances and needs for managing and using the data can be vetted among Task Force members who provide collective experience and technical expertise to problem solving.

Lessons Learned

FESTF has learned many valuable lessons along the way, and its learnings are reflected in the flexible design and data advances that have continued to build over time. At the present time, great emphasis in the species evaluation process is being placed on the comparison of spatial datasets and modeling interactions using spatial relationships combined with product properties evaluated in toxicological and environmental fate studies. Spatial data and environmental modeling can be adequate but not sufficient. For example, spatial data of adequate resolution are available for less than half of the ESA-listed species. Species ranges are broadly described and usually not tied to occupied habitat or even the critical components that must be in an area for the species to survive. Crop spatial data, particularly on minor crops, are not always well characterized. Spatial data do not always accurately portray a specific crop-type (i.e., distinguishing apples from pears). Crop spatial data, has errors at the pixel level, meaning "wheat" might be identified in a county that has none but does have patchy grass. It is complex to spatially represent and evaluate linear use sites such as powerlines, roads, and railroad tracks. It is also complex and challenging to represent pesticide use sites that have no spatial data available (such as residential turf) in the current process that relies so heavily on spatial data. Yet, agencies managing these resources have regulations requiring, among other things, pest control for fire suppression and maintaining visibility.

FESTF has learned that all needs are important to consider in the species assessment process. Not all these needs can be quantified by data aggregation or data management, but the better job FESTF does at that, the more likely the FIFRA/ESA evaluation process is to eventually mature into a stable and predictable regulatory platform.

References

1. TVA, 1978. Tennessee Valley Authority v. Hiram Hill et al., or TVA v. Hill, 437 US 153 (1978).
2. FESTF, 1996. FESTF Feasibility Project Report June 28, 1996 (unpublished report submitted to the Office of Pesticide Programs, EPA).
3. EPA, 2000. USEPA Pesticide Registration Notice 2000-2 (2000). The FIFRA (Federal Insecticide, Fungicide, and Rodenticide Act) Endangered Species Task Force. April 17, 2000.
4. USDOE, 2021. US Department of Energy. Directives Program, Office of Management. Website section titled "Definitions." https://www.directives.doe.gov/terms_definitions/cooperative-research-and-development-agreement-crada (accessed 2021-03-08)
5. EPA, et al., 1980. Interagency agreement between EPA/OPP, US Department of Interior (USDI), Office of Endangered Species (OES), and US Department of Commerce (USDC), National Marine Fisheries Service (NMFS), 1980, as referenced in USEPA OPP EPA-540/9-85-001 Hazard Evaluation Division Standard Evaluation Procedure (for) Ecological Risk Assessment. June 1986.
6. EPA, 2001 and 2005, respectively, for atrazine and 2,4-D. EPA OPP Data Evaluation Record for confidential reports submitted by registrants Syngenta and the 2,4-D Task Force.

Chapter 8

Pollinator Research Task Force—Contributing to the Science of Pollinator Risk Assessments for Pesticides

Joseph D. Wisk*,[1] and Daniel R. Schmehl[2]

[1]BASF Corporation, Agricultural Solution, Global Ecotoxicology Department, 26 Davis Drive, Research Triangle Park, North Carolina 27709, United States

[2]Bayer CropScience LP, Research and Development, 700 Chesterfield Parkway West, Chesterfield, Missouri 63017, United States

***Email: joseph.wisk@basf.com**

In 2012, North America regulatory agencies published a new framework for pollinator risk assessments for pesticides and in 2014, the agencies published their final guidance for assessing pesticide risk to bees. This resulted in enhanced pollinator regulatory testing and science-based risk assessments for pesticide products in North America. However, scientists from both the regulatory and regulated communities recognized several uncertainties in the new guidance including, estimation of residue levels in pollen and nectar of a treated crop, pollen and nectar consumption estimates, the relevance of routes of exposure other than direct contact or through the pollen and nectar diet, and the use of the honey bee as a surrogate for native bee species. In addition, the guidance contained new data requirements, including some studies without fully validated protocols.

To help address these uncertainties, a group of crop protection companies established the Pollinator Research Task Force (PRTF). The mission of the PRTF is to review existing knowledge and develop new data to identify and fulfill generic data gaps, improve and validate test methods, and improve the risk assessment process for insect pollinators. To date, the PRTF has conducted a diverse series of projects covering laboratory test design, field design and population modeling, exposure assessment and toxicity evaluation. PRTF work products are submitted to the United States Environmental Protection Agency (US EPA) and, as appropriate, submitted for publication in scientific journals.

Introduction and Background

There has been significant advancement over the past decade in the pesticide risk assessment process to help ensure that registered uses of crop protection products do not present an unreasonable risk to honey bees and other insect pollinators. The North American regulatory agencies (United States Environmental Protection Agency (US EPA), Health Canada Pest Management Regulatory Agency (PMRA), and California Department of Pesticide Regulation) published a new framework for pollinator risk assessment for pesticides (*1*) in 2012 and final guidance for assessing pesticide risk to bees (2) in 2014. The new guidance, which relies on the European honey bee *(Apis mellifera* L.) as a surrogate species for insect pollinators, includes new testing requirements for effects on both adult and larval honey bees as well as exposure estimates for a screening-level risk assessment. These exposure estimates are incorporated into U.S. EPA's Bee-REX screening-level risk assessment tool (3). The guidance also outlines steps to refine the screening-level risk assessment through the conduct of higher-tier semi-field and field studies to measure exposure to and/or effects on honey bee colonies.

Although the new guidance represents a significant advancement in risk assessment for insect pollinators in North America, there are several uncertainties in the guidance that may lead to over- or under-estimation of exposure and risk. The uncertainties include, estimates of concentrations of pesticide residues in pollen and nectar of a treated crop, pollen and nectar consumption estimates for adult and larval honey bees, the relevance of routes of exposure other than direct contact or through the pollen and nectar diet, and the use of the honey bee as a surrogate for native bee species. In addition, the guidance contains new data requirements, including studies that did not have fully validated test protocols at the time when the guidance was released.

To help address these uncertainties, a group of crop protection companies established the Pollinator Research Task Force (PRTF). The mission of the PRTF is to review existing knowledge and develop new data to identify and fulfill generic data gaps, improve and validate test methods, and improve the risk assessment process for insect pollinators.

As of the beginning of 2021, the PRTF consisted of eight member companies, who are identified in Table 1.

Table 1. List of PRTF Member Companies

BASF Corporation
Bayer CropScience LP
Corteva Agriscience
FMC Corporation
Mitsui Chemicals Agro, Inc.
Syngenta Crop Protection, LLC
UPL NA Inc.
Valent U.S.A. LLC

Project Selection and Work Products

After the Task Force was established, the PRTF held an initial meeting with U.S. EPA Office of Pesticide Programs (OPP) staff; both OPP management staff and members of the Environmental

Fate and Effects Division's Pollinator Team. The purpose of these meetings was to discuss potential projects that would be useful to EPA for informing their risk assessment process for insect pollinators. These meetings have since been held on a yearly basis to review the progress on specific projects and to solicit ideas for new projects. More recently, meetings have been held with Health Canada's PMRA to provide updates and to solicit project ideas that might benefit the risk assessment process of the PMRA.

PRTF projects can be divided broadly into three areas of research: 1) laboratory studies and endpoints, 2) exposure assessment, and 3) field study design and population modeling. PRTF projects are typically highly collaborative, and can involve contract research organizations, consultants, academia, government, and industry. A good example of this was the ring test for a repeated-dose (chronic) toxicity study with honey bee larvae. The ring test involved 15 laboratories across three continents and included contract research, government, industry, and university laboratories.

All PRTF work products are submitted to US EPA and if appropriate, submitted for publication in scientific journals. There have also been numerous presentations of PRTF projects presented at scientific conferences. Work products are also posted on the PRTF website (PollinatorResearchTaskForce.com). A brief description of each PRTF project to date is summarized in the following sections.

Table 2. PRTF Projects on Laboratory Studies and Endpoints

Project	*Status as of March 2021*
Validation and Ring Testing of Honey Bee Chronic Larval Study	Complete
Comparison of the Relative Toxicity of Technical Grade Active Ingredients Versus Formulated End-use Products to Adult Honey Bees	Complete
Investigation of Alternative Exposure Vehicles for Honey Bee Chronic Laboratory Studies	In progress
Standardization of the Study Design for the Honey Bee Toxicity of Residues on Foliage (RT25) Test	In progress
Relevance of Weight of Emerged Bees as a Regulatory Endpoint in the Honey Bee Chronic Larval Toxicity Study	In progress
Comparison of Honey Bee Larval Acute and Chronic Endpoints for Redundancy and Relevance to Risk Assessment	In progress

Laboratory Studies and Endpoints

The North American regulatory authorities rely upon laboratory-based studies for the generation of pesticide toxicity data on individual honey bees that can also be used as a surrogate for other insect pollinators (2). The laboratory toxicity data are paired with default or empirical exposure estimates that inform a screening-level risk characterization that represents a worst-case exposure scenario. There were technical challenges with test performance at the time the North American bee guidance was released in 2014 (2), particularly for larval honey bee assays and the toxicity of residues on foliage (RT_{25}) assay. There were also limited data on the use of solvents and emulsifiers as carriers for test substances in laboratory tests when a pesticide is otherwise not homogeneous or dissolved within the bee diet. Finally, there is uncertainty on how sublethal endpoints that are recorded in laboratory studies may translate to assessment endpoints (*i.e.* survival, growth, or reproduction), particularly how the weight of eclosed adults relates to the assessment endpoints within the honey bee chronic larval assay. A list of PRTF projects involving laboratory studies and endpoints are found in Table 2, followed by detailed project descriptions in the paragraphs that follow.

Validation and Ring Testing of Honey Bee Chronic Larval Study

One of the first projects that PRTF worked on was to organize a formal ring testing of a protocol for a chronic (repeated-dose) toxicity test with honey bee larvae. The project was initiated at the request of the US EPA.

A chronic larval toxicity study is a new data requirement in the North American guidance for assessing pesticide risk to bees (2). The study involves four consecutive days of dosing larvae during the 6-day larval stage, followed by observation and evaluation of survival through pupation and emergence of adult honey bees. However, laboratories were experiencing inconsistent performance in meeting the control survival validity criteria when using a method based on an Organization for Economic Cooperation and Development (OECD) draft guidance document (4). The PRTF refined the test protocol by incorporating methodology that was developed at the University of Florida (5) into the existing method. Protocol refinements include changes to the composition of the larval diet, source of royal jelly and some modifications to the rearing environment.

The formal ring test was conducted at 15 laboratories around the world using the toxic reference standard dimethoate. Participating laboratories included two industry laboratories (BASF SE, Limburgerhof, Germany and Bayer CropScience, Monheim, Germany), nine contract research organizations (Biochem Agar, GmbH, Machern, Germany; Eco-Saf Technology, Beijing, China; Eurofins Agro, Jena, Germany; Eurofins Agriscience Service, LLC, Mebane, NC, USA; IBACON GmbH, Roβdorf, Germany; Innovative Environmental Services [IES], Ltd., Witerswil, Switzerland; Smithers, LLC, Snow Camp, NC, USA; SynTech Research, Inc., Witerswil, Switzerland, and SynTech Research, Inc., Stilwell, KS, USA), three academic laboratories (Michigan State University, East Lansing, MI, USA; Nanjing Institute of Environmental Sciences, Nanjing, China and University of Florida, Gainesville, FL, USA), and one government laboratory (LAVES Niedersachsen, Wardenburg, Germany).

Results from the ring test demonstrated that:

- Acceptable control survival was achieved in 87% of the tests.
- Study endpoints with dimethoate as a reference toxicant (survival and emergence) only varied within a factor of 2X, indicating good test performance.
- One additional finding was that acetone, which is the recommended vehicle for introducing the test substance into the test diet, was toxic to honey bee larvae at concentrations $\geq 2\%$.

The modified method is now being used by testing facilities and the performance of this test has become much more reliable. The validation report (6) is posted on the PRTF website (PollinatorResearchTaskForce.com).

Comparison of the Relative Toxicity of Technical Grade Active Ingredients versus Formulated End-Use Products to Adult Honey Bees

In North America, laboratory testing conducted to evaluate the potential effects of pesticides on individual honey bees utilizes technical grade active ingredient (TGAI). However the active ingredient is formulated as a Typical End-use Product (TEP) when commercially sold and used in the field. The TEP is generally used in higher-tier (semi-field/field) studies when required by regulatory authorities based upon the laboratory toxicity data on individual bees. There is uncertainty of whether results generated in the laboratory with the TGAI truly represent the potential toxicity of the end-use product to pollinators.

The PRTF developed and evaluated a database of laboratory studies conducted with both types of test substances to evaluate the relative toxicity of TGAIs versus TEP. The database contained acute oral and contact data on adult honey bees for approximately 150 different chemicals and associated formulated products. The chemicals were classified according to US EPA's toxicity classification categories of highly toxic ($LD_{50} < 2$ µg a.i./bee), moderately toxic ($10.9 > LD_{50} > 2$ µg a.i./bee), and practically non-toxic ($LD_{50} \geq 11$ µg a.i./bee). The toxicity classification for each chemical was compared to determine whether differences in toxicity were observed between the TGAI and TEP.

Results from the comparison of acute contact toxicity endpoints demonstrated:

- TGAI and TEP LD_{50} values were in the same toxicity category 94.5% of the time.
- The TGAI was more toxic 4.1% of the time.
- The TEP was more toxic 1.4% of the time.

Results from the comparison of acute oral toxicity endpoints demonstrated:

- TGAI and TEP LD_{50} values were in the same toxicity category 90.3% of the time.
- The TGAI was more toxic 4.5% of the time.
- The TEP was more toxic 5.2% of the time.

The results of this comparison indicated no patterns of greater toxicity being observed from the formulated end-use product in relation to the TGAI. These results suggest that a reliance on the TGAI in laboratory testing is adequate and protective of honey bees for use in screening-level risk assessments. This review has been published in the *Journal of Economic Entomology* (7).

Investigating Alternative Exposure Vehicles for Honey Bee Chronic Laboratory Studies

The PRTF initiated a project to investigate alternative exposure vehicles for use in both honey bee larval and adult chronic tests based on the results from the honey bee chronic larval validation and ring test project which showed that a 2% concentration of acetone (currently identified as permissible in the OECD draft guidance document (4)) were toxic to honey bee larvae. Having effective and potentially non-toxic solvent vehicles to deliver test substances to the test diets in these tests is particularly important for test substances that have low aqueous solubility and low toxicity, and would typically require a higher solvent load to achieve higher concentrations of the test substance in the test diets. The PRTF project builds on the work of Bluhm et al. (2017) with sodium lignosulfonate and xanthan (8). The purpose of the project is not to identify a single best-possible solvent, but to identify several options that are acceptable to regulatory authorities for use in bee testing.

Initial pilot trials for the project were initiated in 2018, followed by the main experimental phase of the project in 2019-2020. There are six participating laboratories; four from Europe (SynTech Research, Inc., Witerswil, Switzerland; Biochem Agar GmbH, Machern, Germany; Innovative Environmental Services [IES], Ltd., Witerswil, Switzerland and Eurofins Agro, Jena, Germany) and three from the U.S. (Smithers, North Carolina; Eurofins Agriscience Service, LLC, Mebane, NC and Bayer CropScience, Chesterfield, MO). Six candidate vehicles are being evaluated: ethanol, n-propanol, isopropanol, propylene glycol, triethlyene glycol and xanthan.

The analysis of the experimental work is expected to be completed in 2021 and the results, along with recommendations will be submitted for publication.

Standardization of the Study Design for the Honey Bee Toxicity of Residues on Foliage (RT25) Test

The honey bee toxicity of residues on foliage, or RT_{25} test, is a US EPA conditional data requirement. The test is required when the acute contact LD_{50} from the required adult honey bee contact toxicity test is < 11 µg a.i./bee. The test is intended to estimate the time that a pesticide remains toxic to bees on treated foliage in the field. The US EPA uses this information for bee hazard and risk mitigation language on pesticide labels, *i.e.*, to determine restrictions around at-bloom applications. In addition to the U.S., this study has traditionally been required in South Korea and more recently has become a conditional requirement in Brazil.

A problem with this test has been inconsistencies in the generated data and the data not always being predictive of toxicity in the field. Because of these inconsistencies, the U.S. EPA requested that the PRTF evaluate and validate the experimental design for this study.

A review of the test guideline by the PRTF identified several potential sources of variability that could result in inconsistent results between test facilities. The PRTF developed a test protocol which required more consistent test parameters, including standardizing the exposure parameters by requiring test cages of similar size and dimensions, and optimizing the placement of treated foliage in the test cages. The revised protocol also specifies the use of lab-reared worker honey bees of similar age (3-5 d), field grown alfalfa as the test crop, and spray application and weathering in the field, all aimed at increasing the predictive value of study results.

Trials were conducted at three test facilities during 2020: Eurofins, Mebane, NC; Smithers, Snow Camp, NC and Eurofins, Brazil. Additional evaluation on the field application phase of the study design will occur in 2021 based on the results from 2020, which continued to show inconsistencies. Reports from the 2020 trials and future phases of the program will be submitted to U.S. EPA and posted on the PRTF website.

Relevance of Weight of Emerged Bees as a Regulatory Endpoint in the Honey Bee Chronic Larval Toxicity Study

US EPA has requested registrants to include an assessment of weights of emerged bees as a study endpoint in the 22-day larval chronic study design despite no current requirement in the OECD guidance document (4). The PRTF initiated a project to evaluate the relevance of this sublethal observation for assessment endpoints (*i.e.* growth) in regulatory decision making. To perform this analysis, assay data were compiled, and a meta-analysis was conducted.

During 2019, the PRTF organized an analysis of data from 26 studies that had been conducted by PRTF member companies. In 2020, the analysis was expanded to include data from 20 additional studies conducted by non-PRTF member companies. Of the studies analyzed, 48% were conducted with herbicides, 26% with fungicides and 26% with insecticides/acaricides.

Preliminary results indicate:

- The weight at emergence endpoint resulted in a lower study endpoint in 17% of the evaluated studies (8/46).
- For the studies where weight at emergence was the lowest statistically significant endpoint, the weights were only 4-10% lower than the control and may not result in biological relevance.

Once the analysis is complete, a manuscript of this work will be submitted for publication.

Comparison on Honey Bee Larval Acute and Chronic Endpoints for Redundancy and Relevance to Risk Assessment

According to the North American guidance (2), both a 7/8-day single-dose (acute) and 22-day repeated-dose (chronic) studies with honey bee larvae are tier 1 data requirements. In both studies, an LD_{50} for larvae can be determined; based on a single day exposure and a cumulative dose. The objective for this project was to compare the day-7/8 LD_{50}/LC_{50} survival endpoints from both larval studies with the same test substances to determine if the acute study provides any additional value to the risk assessment over having only the chronic study. The chronic larval study (see previous section, *Validation and ring testing of honey bee chronic larval study*) provides endpoints covering all stages of brood development including those obtained from the acute study. Now that the protocol from this study has been validated and ring-tested, the older, single-dose acute larval study has become redundant.

The PRTF has evaluated the results from studies on 43 active ingredients, including 10 herbicides, 17 fungicides, 15 insecticides and 1 herbicide safener. In this project, the day- 7/8 LD_{50}/LC_{50} value from each study was considered different when they were at least 2-fold lower or when a definitive value was lower than a non-definitive (*i.e.*, a greater than) value. The results of the analysis are found in Table 3.

Table 3. Comparison of Endpoints between Single- and Repeat-Dose Honey Bee Larvae Toxicity Studies

Summary of Results	*7/8-day LD_{50}*	*$LD_{50}/4$*	*7/8-day LC_{50}*
Single-dose study lower	16% (7 of 43)	0% (0 of 43)	0% (0 of 41)
Repeat-dose study lower	16% (7 of 43)	40% (17 of 43)	36% (15 of 41)
Both studies similar	67% (29 of 43)	60% (26 of 43)	63% (26 of 41)

The results of this analysis demonstrated:

- 7/8-day larval LD_{50} values are similar for both study designs even when based on a single (daily) or cumulative (over 4 days) dose.
- When converting the cumulative dose to a daily dose (by dividing the LD_{50} by 4, which is the current US EPA policy), results from the cumulative-dose studies are often lower than those from the single-dose study.

- Similarly, when endpoints are based on the dietary concentration (LC_{50}) the values are lower or similar for the repeat-dose studies.

Overall, in most of the cases, the repeat-dose larval study provides the most conservative 7/8-day LD_{50}/LC_{50} values for use in a pollinator risk assessment. In addition, because developing honey bee larvae will not likely be exposed to a single daily dose of a pesticide within a colony, the results from the single-dose study does not provide any additional information to inform a risk assessment. Therefore, the PRTF believes that the single-dose honey bee larval study is redundant and can be eliminated to reduce unnecessary animal testing. A manuscript for publication of this project is in preparation.

Exposure Assessment

The North American regulatory authorities rely upon default or empirical exposure estimates that inform a screening-level risk characterization that represents a worst-case exposure scenario in the current risk assessment framework. The dietary exposure (pollen/nectar) defaults are part of the Bee-REX model (3), but there are uncertainties related to consumption estimates and the appropriate dietary matrices for both honey bees and other insect pollinators. Additionally, there is concern that the current default Bee-REX exposure estimates are based upon limited data that may not be directly relevant to bee dietary matrices. A list of PRTF projects involving exposure assessments are found in Table 4, followed by detailed project descriptions in the following paragraphs.

Table 4. PRTF Projects Related to Insect Pollinator Exposure Assessments

Project	*Status as of March 2021*
Literature Review of Individual Nectar-Foraging Honey Bees and Whole Colony Consumption of Nectar and Pollen	Complete
Probabilistic Modeling of Nectar Consumption of Foraging Boney Bees	Complete
Relevance of Plant Guttation Water as an Exposure Route for Honey Bees	Complete
Organization and Participation in Non-*Apis* Bee Exposure Workshop	Complete
Compilation of Data from Pollen and Nectar Residue Studies to Refine Bee-REX Tier 1 Estimates	In progress
Evaluation of the Briggs Plant Uptake Model for Prediction of Pesticide Residues in Pollen and Nectar After Soil Applications of Systemic Compounds	In progress

Literature Review of Individual Nectar-Foraging Honey Bees and Whole Colony Consumption of Nectar and Pollen

One of the key factors in predicting the dietary exposure of pesticides by honey bees is estimating the amount of pollen and nectar that honey bees consume. To assess dietary exposure, the North American regulatory agencies modeled daily pollen and nectar consumption by different castes of honey bees and presented this work in their framework for pollinator risk assessments (*1*). However, the agencies did not provide any estimates of pollen and nectar being consumed by whole colonies, which is a major factor in predicting exposure to, and effects of, pesticides to honey bee colonies under field conditions.

Therefore, the PRTF initiated a comprehensive literature review on the nutrition and energetics of honey bees to collect available data on 1) the quantity of nectar required by nectar foragers and 2) the quantities of nectar, pollen and supplements that are consumed by honey bee colonies. Several studies on the energetic requirements of individual nectar foragers were found that provided additional information such as:

- Estimating the sugar requirements for foragers under different conditions (e.g., flying vs. resting).
- The amount of time foragers spend outside the hive.
- There was limited information on food consumption by whole colonies.

This information can be used to model nectar consumption rates of foraging honey bees. Further research is needed to predict whole colony food consumption. The results of this review have been published in the Journal *Apidologie* (*9*).

Probabilistic Modeling of Nectar Consumption of Foraging Honey Bees

Based on the North American regulatory agencies' analysis, as presented in their framework for pollinator risk assessments (*1*), the highest consumption for any honey bee caste was for nectar foragers, with a median ingestion rate of 292 mg nectar/bee/day. This value is used in the agencies' Bee-REX model for estimating dietary exposure in screening-level risk assessments (3). However, the agencies' model was based on limited input data and was not subjected to a sensitivity analysis.

To evaluate the agencies' model, the PRTF initiated a project to use the data from the comprehensive literature review of individual nectar-foraging honey bees and whole colony consumption of nectar and pollen to parameterize a probabilistic model to estimate nectar requirements for nectar-foraging honey bees. Model input data included the number of trips per day, trip duration, fraction of the trip spent flying, sugar requirement for flight, sugar requirement to meet resting metabolic rate, sugar content of nectar, and maximum time the foragers were outside the hive.

Based on a probabilistic assessment using the revised inputs:

- A median ingestion rate of 44 mg nectar/bee/day was estimated for nectar-foraging honey bees.
- These results indicate that estimates of dietary consumption by nectar-foraging honey bees may be a significant source of conservativism in the North American agencies' screening-level risk assessment.

The details and results of this work have been published in the Journal *Apidologie* (*10*).

Relevance of Plant Guttation Water as a Potential Exposure Route for Honey Bees

Guttation is a process of exudation of xylem sap on the edges of plant leaves. This process occurs in vascular plants when there is high soil moisture and the plant then has excess water. Guttation fluid may possibly be used as a water source by honey bees and there is a potential for this fluid to contain pesticide residues. Some stakeholders have raised concerns that current risk assessment processes do not take into account potential exposure by way of guttation water. The PRTF commissioned a review of recent literature on this topic and 13 relevant papers were found. All of the studies involved crops treated neonicotinoid insecticides and nearly all were seed treatment uses. There was one study with a drench application, one with an in-furrow application and one where the compound was applied as a spray to turf grass and watered in. Crops evaluated included corn, oilseed rape, cantaloupe, winter cereal crops, sugar beets and turf grass.

The results and conclusions of this review are:

- Residue levels in guttation water vary between crops, but generally decline as plants develop. The highest residues were found in corn guttation water.
- There is limited collection of guttation water by honey bees in the field, reducing concern for exposure by this potential route.
- There are no effects of guttation water exposure on honey bee colonies observed in the field.
- Under certain circumstances, residues of pesticides in guttation water may reach levels that are potentially toxic to bees, but only if they were to consume it in sufficient quantities to reach a toxic dose.
- The overall conclusion from this literature review was that guttation water is not an important route of pesticide exposure to honey bees.

This review has been published in the Journal *Apidologie* (*11*).

Non-Apis Bee Exposure Workshop

In January 2017, the US EPA hosted a multi-stakeholder workshop with the goal of understanding potential differences in pesticide exposure of honey bees versus native or non-*Apis* bees. The PRTF contributed to the workshop through funding and participation on the organizing committee and in the workshop.

The workshop participants identified potential routes of exposure that are unique to non-*Apis* bees versus honey bees and identified research needs. Bee taxa evaluated were bumble bees, stingless bees, and solitary bees. The overall conclusions of the workshop were:

- Dietary exposure through consumption of pollen and nectar was an important exposure route discussed and received the highest risk exposure scoring across all bee species.
- Currently, only limited information is available on average provision composition and consumption rates of non-*Apis* bee species, which limits the ability to quantify dietary exposure.
- Honeydew, water, and guttation fluids were evaluated but were considered minor routes of exposure.
- Several data gaps were identified that hindered the full quantification of exposure to non-*Apis* bees, especially when these bees are exposed via nesting materials such as soil (*e.g.*, blue orchard bees and alkali bees), leaves (*e.g.*, leafcutter bees) or a combination of soil and leaves (*e.g.*, stingless bees).
- Basic conceptual models and preliminary exposure equations were discussed that would help to address quantification of these exposure routes in the future and, subsequently, would allow comparison with honey bee exposure estimates.

The workshop proceedings are summarized in six publications in the journal *Environmental Entomology* (*12–17*).

Compilation of Pollen and Nectar Residue Data to Refine Bee-REX Tier 1 Estimates

Another critical component of the dietary risk assessment for honey bees and other insect pollinators is the prediction of residue levels in pollen and nectar of a treated crop. When the North American regulatory agencies developed their framework for assessing the risk of pesticides to bees (*1*), there was a limited number of studies available to develop a robust quantitative relationship or nomogram for predicting residues in these two matrices. Based on this limited data set, the agencies used the long grass component of EPA's Terrestrial Residue Exposure Model (T-REX) (*18*) as the screening-level estimate for calculating a Residue Unit Dose (RUD) for pollen and nectar after a direct application to blooming crops. This long grass estimate has been incorporated into the agencies' Bee-REX model (3) as a surrogate for both pollen and nectar.

However, since the publication of the North American agencies' framework and final guidance (2), hundreds of pollen and nectar residues studies have been conducted and submitted to regulatory authorities around the world. For example, the US EPA already has an extensive dataset for four neonicotinoid insecticides (imidacloprid, clothianidin, thiamethoxam, and dinotefuran), as well as sulfoxaflor. The European Food Safety Authority (EFSA) is already using a pollen and nectar residue database to derive residue RUD estimates. In addition, both the International Committee for Plant Bee Relationships (ICPBR) and the OECD have identified a global pollen and nectar residue database as a desired tool.

The PRTF initiated a project to merge the EFSA and EPA (including the four neonicotinoids and sulfoxaflor) databases and assemble the data in a spreadsheet format to provide the US EPA. The available information in the combined database will likely be sufficient to calculate pollen and nectar RUDs for several crop groups, application types, exposure duration, and possibly other categories. The PRTF anticipates the delivery of the database to EPA during the first half of 2021.

Evaluation of Briggs Plant Uptake Model

A potential route of exposure that is unique for systemic pesticides is through uptake into plants and translocation into pollen and nectar after soil applications or during use as a seed treatment. The Briggs's Plant Uptake Model is incorporated into the Bee-REX screening-level tool (3) for screening-level risk assessments to estimate potential translocation into these bee-relevant matrices after soil applications. The baseline Brigg's model was based on research measuring the translocation of non-ionized chemicals from soil into the barley stems (*19*, *20*). The baseline model was further modified based on work of Ryan et. al. (1988) (*21*).

However, the Briggs's model was developed to predict residues translocating into stems (not pollen or nectar) and has only been validated in two crops. Therefore, the PRTF initiated a project in 2020 to gather information on pollen and nectar residue studies with soil-applied pesticides from both the literature and from registrant-generated studies. The project will also include a literature search for any models that have been developed since the North American agencies' framework has been developed. The goal is to compare actual data with the Briggs's model predictions and determine the appropriateness of the model or if a more suitable model is available. It is expected that this project will continue through at least the end of 2021.

Field Study Design and Population Modeling

One of the risk assessment components listed in the North American regulatory agencies' final guidance for assessing pesticide risk to bees (2) was the development and future use of models to predict effects on honey bee colonies. Based on this recommendation, the PRTF initiated a project using the mechanistic honey bee colony model BEEHAVE (22). BEEHAVE was designed to model the influence of various stressors, as well as pollen and nectar-foraging behavior in different landscapes, on the development of honey bee colonies.

The first goal of this project was to assess BEEHAVE as a tool to model large-scale colony feeding studies (LSCFS) and their outcomes in a specific region of the U.S. (i.e., North Carolina). A second goal was to use BEEHAVE to predict overwintering survival of LSCFS control colonies, depending on external conditions and study designs, including: initial colony conditions, beekeeper interventions, resource availability (spatially and temporally) to the colony and weather conditions.

In this project, BEEHAVE was calibrated using control colony data from available LSCFS. Both qualitative (visual) and quantitative (goodness-of-fit) validation criteria were developed and applied. The criteria accounted for within apiary variability and measurement error in the study data. The validation, which focused on adult bees and honey stores suggested a good model performance prior to overwintering.

BEEHAVE was used to determine which factors lead to weaker or stronger colonies in the LSCFS. It was determined that initial colony condition and feeding pattern were important for the condition of the colony in autumn. It was determined that the tested landscapes and weather had negligible impact.

Based on these findings, the recommendations for future LSCFS test designs were made:

- During colony set-up, researchers should focus on initial honey stores, which are most indicative of autumn colony size and honey stores. Colonies with small initial honey stores of 4 kg or less should be excluded from the study.
- Colonies should be provided with more sugar in late summer and early autumn to adequately grow the colonies prior to winter. Also, to increase the population of adult bees in the colony during autumn, supplemental feeding should begin earlier. However, if the goal is to grow autumn honey stores, supplemental feeding should begin later.

Two manuscripts of this work have been published in the *Journal of Environmental Toxicology and Chemistry* (23, 24).

Summary and Conclusions

The PRTF continues to conduct projects to attempt to address uncertainties in the North American risk assessment process for pesticides and insect pollinators. The goal is to provide information to inform the scientifically robust framework that has been developed and is being used by regulatory authorities. The success of this Task Force has been due to the generous funding by member companies and the high level of collaboration between company scientists and other project partners from universities, consulting firms and contract research organizations. Regular consultation with North American regulatory authorities has been a critical factor for determining projects to work on and another reason for this Task Force achievements. To date, the work of the PRTF has led to 12 scientific journal publications, 7 formal reports submitted to US EPA and numerous presentations of work at scientific conferences. Information on the PRTF, including access to work products and publications can be found at: PollinatorResearchTaskForce.com.

Acknowledgments

The generation of the information and data presented here was funded completely or in-part by the Pollinator Research Task Force. In addition to the authors, the following individuals contributed to the projects presented here and/or the review of this manuscript: Rajwinder Singh (BASF Corporation), David Fischer, Ana Cabrera, Tim Fedricks, Allen Olmstead (Bayer CropScience LP), Vince Kramer, Bridget O'Neill, Verissimo Sa (Corteva Agriscience), Bibek Sharma, Tiffany Carro (FMC Corporation), Tim Joseph (Landis International on behalf of Mitsui Chemicals Agro, Inc.), Max Feken, Nika Galic, Jay Overmyer (Syngenta Crop Protection, LLC), Crystal Layton, Stephanie Tokar (UPL NA, Inc.), and Silvia Hinarejos (Sumitomo Chemical Company, Ltd. and Valent U.S.A., LLC).

References

1. White Paper in Support of the Proposed Risk Assessment Process for Bees 2012. U.S. Environmental Protection Agency Office of Pesticide Programs Environmental Fate and Effects Division, Health Canada Pest Management Regulatory Agency Environmental Assessment Division, and California Department of Pesticide Regulation. Submitted to the FIFRA Scientific Advisory Panel, September 11–14, 2012.
2. Guidance for Assessing Pesticide Risk to Bees 2014. U.S. Environmental Protection Agency Office of Pesticide Programs, Health Canada Pest Management Regulatory Agency, and California Department of Pesticide Regulation.
3. Models for Pesticide Risk Assessment, Bee-REX Version 1.0, 2015. U.S. Environmental Protection Agency Washington, D.C., https://www.epa.gov/sites/production/files/2015-11/beerexv1.0.xlsx (accessed February 1, 2021).
4. OECD Draft Guidance Document, Honey Bee (*Apis mellifera*) Larval Toxicity Test, Repeated Exposure, 20 July 2015. Organization for Economic Cooperation and Development, Paris, France.
5. Schmehl, D. R.; Tomé, H. V. V.; Mortensen, A. N.; Ferreira Martins, G.; Ellis, J. D. Protocol for the *in vitro* Rearing of Honey Bee (*Apis mellifera*) Workers. *J. Apicultural Res.* **2015**, 55 (2), 113–129.
6. Clark, S. L. Results of the Honey Bee (Apis mellifera L.) Toxicity Following Chronic (Repeated) Exposure International Ring Test Performance. Technical report prepared for the Pollinator Research Task Force, 2017. Report available at: http://pollinatorresearchtaskforce.com/wp-content/uploads/2019/03/PRTF-Projects-RingStudy-Report-Feb-2017.pdf (accessed February 2, 2021).
7. Spurill, S. E.; O'Neill, B. F.; Hinarejos, S.; Cabrera, A. R. A Comparison of Acute Toxicity Endpoints for Adult Honey Bees for Technical Grade Active Ingredients and Typical End-use Products as Test Substance. *J. Econ. Entomol.* **2020**, *113* (2), 1015–1017.
8. Bluhm, W.; Gladbach, D. J.; Kling, A.; Ruhland, S.; Sack, D.; Schmitzer, S.; Theis, M.; Wollert, S.; Hanewald, N. Limited Solubility of Test Items in Regulatory Honey Bee (Apis mellifera) Testing: Potential use of Solvents, Solubilizers and Viscosifiers with Aqueous Sugar Solution, Presented at SETAC Europe, May, 2017, Brussels, Belgium.
9. Rodney, S.; Purdy, J. Dietary Requirements of Individual Nectar, and Colony-Level Pollen and Nectar Consumption: A Review to Support Pesticide Exposure Assessment for Honey Bees. *Apidologie* **2020**, *51*, 163–179.

10. Rodney, S.; Kramer, V. Probabilist Assessment of Nectar Requirements for Nectar Foraging Honey Bees. *Apidologie* **2019**, *51*, 180–200.
11. Schmolke, A.; Kearns, B.; O'Neill, B. Plant Guttation Water as a Potential Route for Pesticide Exposure in Honey Bees: A Review of Recent Literature. *Apidologie* **2018**, *49*, 637–646.
12. Bireley, R.; Borges, S.; Cham, K.; Epstein, D.; Garber, K.; Hart, C.; Hou, W.; Ippolito, A.; Pistorius, J.; Poulsen, V.; Sappington, K.; Steeger, T. Preface: Workshop on Pesticide Exposure Assessment Paradigm for Non-*Apis* Bees. *Environ. Entomol.* **2018**, *48* (1), 1–3.
13. Boyle, N. K.; Pitts-Singer, T. L.; Abbott, J.; Alix, A.; Cox-Foster, D. L.; Hinarejos, S.; Lehmann, D. M.; Morandin, L.; O'Neill, B.; Raine, N. E.; Singh, R.; Thompson, H. M.; Williams, N. N.; Steeger, T. Workshop on Pesticide Exposure Assessment for *Non-Apis* Bees: Foundation and Summaries. *Environ. Entomol.* **2018**, *48* (1), 4–11.
14. Gradish, A. E.; van der Steen, J.; Scott-Dupree, C. D.; Cabrera, A. R.; Cutler, G. C.; Goulson, D.; Klein, O.; Lehman, D. M.; Lückmann, J.; O'Neill, B.; Raine, N. E.; Sharma, B.; Thompson, H. Comparison of Pesticide Exposure in Honey Bees (Hymenoptera: Apidae) and Bumble Bees (Hymenoptera: Apidae): Implications for Risk Assessment. *Environ. Entomol.* **2018**, *48* (1), 12–21.
15. Sgolastra, F.; Hinarejos, S.; Pitts-Singer, T. L.; Boyle, N. K.; Joseph, T.; Lückmann, J.; Raine, N. E.; Singh, R.; Williams, N. M.; Borsh, J. Pesticide Exposure Assessment Paradigm for Solitary Bees. *Environ. Entomol.* **2018**, *48* (1), 22–35.
16. Cham, K. O.; Nocelli, R. C. F.; Borges, L. O.; Viana-Silva, F. E. C.; Tonelli, C. A. M.; Malaspina, O.; Menezes, C.; Rosa-Fontana, A. S.; Blochtein, B.; Freitas, B. M.; Pires, C. S. S.; Oliveira, F. F.; Contrera, F. A. L.; Torezani, K. R. S.; de Fátima Riberio, M.; Siqueira, M. A. L.; Rocha, M. C. L. S. A. Pesticide Exposure Assessment Paradigm for Stingless Bees. *Environ. Entomol.* **2018**, *48* (1), 36–48.
17. Hinarejos, S.; Abbott, J.; Alix, A.; Sharma, B.; Cabrera, A.; Joseph, T.; O'Neill, B.; Singh, R.; Thompson, H. Non-*Apis* Bee Exposure Workshop: Industry Participants' View. *Environ. Entomol.* **2018**, *48* (1), 49–52.
18. Models for Pesticide Risk Assessment, Terrestrial Residue Exposure Model (T-REX), Version 1.5.2, 2013. U.S. Environmental Protection Agency, Washington, D.C., https://epa.gov/sites/production/files/2013/T-Rex1.5.2 (accessed Feb 2, 2021).
19. Briggs, G. G.; Bromilow, R. H.; Evans, A. A. Relationships Between Lipophilicity and Root Uptake and Translocation of Non-Ionised Chemicals by Barley. *Pesticide Sci.* **1982**, *13*, 495–504.
20. Briggs, G.; Bromilow, R.; Evans, A.; Williams, M. Relationships Between Lipophilicity and the Distribution of Non-Ionised Chemicals in Barley Shoots Following Uptake by the Roots. *Pesticide Sci.* **1983**, *14*, 492–500.
21. Ryan, J. A.; Bell, R. M.; Davidson, J. M.; O'Connor, G. A. Plant Uptake of Non-Ionic Organic Chemicals from Soils. *Chemosphere* **1988**, *17* (12), 2299–2323.
22. Becher, M. A.; Grimm, V.; Thorbek, P.; Horn, J.; Kennedy, P. J.; Osborne, J. L. BEEHAVE: A Systems Model of Honeybee Colony Dynamics and Foraging to Explore Multifactorial Causes of Colony Failure. *J. Appl. Ecol.* **2014**, *51*, 470–482.

23. Schmolke, A.; Abi-Akar, F.; Roy, C.; Galic, N.; Hinarejos, S. Simulating Honey Bee Large-Scale Colony Feeding Studies Using the BEEHAVE Model – Part I Model Validation. *Environ. Toxicol. Chem.* **2020**, *39* (11), 2269–2285.
24. AbiAkar, F.; Schmolke, A.; Roy, C.; Galic, N.; Hinarejos, S. Simulating Honey Bee Large-Scale Colony Feeding Studies Using the BEEHAVE Model – Part II Analysis of Overwintering Outcomes. *Environ. Toxicol. Chem.* **2020**, *39* (11), 2286–2297.

Chapter 9

Use of CARES NG to Conduct a Cumulative Dietary Exposure Assessment for Chlorotriazine Compounds: A Case Study

Jason Johnston[*,1] **and Mark Breen**[*,2]

[1]**The Acta Group, 2200 Pennsylvania Avenue NW, Suite 100W, Washington, D.C. 20037, United States**

[2]**Creme Global, 4th Floor, The Design Tower, Trinity Technology & Enterprise Campus, Grand Canal Quay, Dublin D02 P956, Ireland**

[*]**Email: jej8641@gmail.com**

[*]**Email: mark.breen@cremeglobal.com**

Estimation of dietary exposures to pesticide residues in agricultural and livestock commodities is critical in regulating the safe use of pesticide products. The U.S. Environmental Protection Agency (EPA) has long used the Dietary Exposure Evaluation Model-Food Commodity Intake Database (DEEM-FCID or DEEM) successfully for such purposes. Increasingly complex assessments may sometimes be desired. A particular example is the cumulative dietary assessment for the chlorotriazines (atrazine and simazine), for which 4-day average dietary exposures are necessary due to the toxicology point of departure (POD) selected for risk characterization. In the 2018 chlorotriazine cumulative assessment, EPA used point estimates for residues and 100% crop treated in combination with 2-day average food consumption data to approximate 4-day average dietary exposures for atrazine and simazine. EPA then summed the two estimates to conservatively approximate the cumulative chlorotriazine 4-day average dietary exposures. The dietary model in the Cumulative and Aggregate Risk Evaluation System–Next Generation (CARES NG) permits a more refined calculation of multi-day rolling average exposures for both single-chemical and cumulative assessments. Atrazine and simazine are both registered for use on field, pop, and sweet corn. EPA's cumulative approach implies that all corn commodities are treated with both compounds. CARES NG permits creation of a residue input file that preserves EPA's assumption of 100% crop treated for corn commodities while accounting for the relative market share of atrazine and simazine applications to corn. The assessments in this case study illustrate the impact of these CARES NG features on chlorotriazine cumulative dietary exposure estimates.

Introduction

The U.S. Environmental Protection Agency (EPA) completed a cumulative human health risk assessment for the chlorotriazine class of pesticide active ingredients in 2018 (*1*). EPA's chlorotriazine cumulative exposure and risk assessment includes the following components:

- 4-day average dietary exposure (food only),
- Background (chronic) dietary exposure (food only),
- Residential post-application exposure,
- 4-day cumulative dietary aggregate (food + drinking water) exposure, and
- 4-day cumulative aggregate (background food + drinking water + residential) exposure.

This case study focuses exclusively on the 4-day average dietary (food only) exposure assessments conducted by EPA using the Dietary Exposure Evaluation Model-Food Commodity Intake Database (DEEM-FCID or DEEM) (*2*). The intent of this case study is to explore options to refine EPA's assessments using the capabilities of the Cumulative and Aggregate Risk Evaluation System–Next Generation (CARES NG) model (*3*).

The chlorotriazine common mechanism group (CMG) includes three active ingredients: atrazine, propazine, and simazine. The chlorinated metabolites of these chemicals are also included in the CMG, but the hydroxytriazine metabolites, which have a separate toxic effect, are not. The toxicological endpoint for the chlorotriazines and chlorinated metabolites is the disruption of the luteinizing hormone surge observed in rat studies. The critical time period to elicit this effect is a 4-day exposure, and the appropriate exposure metric of interest in the dietary assessment is therefore 4-day average exposures, as discussed in EPA guidance on conducting aggregate (*4*) and cumulative (*5*) exposure and risk assessments. EPA selected the benchmark dose (BMD) from an atrazine study conducted in rats as the point of departure (POD) for purposes of characterizing risks. Physiologically based pharmacokinetic (PBPK) modeling was used to convert the atrazine rat POD to a human-relevant internal dose metric, from which pathway- and subpopulation-specific PODs for each of the three chlorotriazines were estimated.

To begin the cumulative dietary exposure assessment process for the chlorotriazines, EPA first conducted single-chemical dietary (food only) assessments for atrazine (*6*) and simazine (*7*). EPA did not estimate propazine dietary exposures because the only commodity with a registered propazine use is grain sorghum, which has no reported dietary consumption. Most chlorotriazine tolerances are based on the sum of the limits of quantification (LOQ) for the residues of concern (i.e., parent and chlorinated metabolites), so EPA estimated anticipated residues (ARs) in agricultural commodities as one-half of the tolerance. Atrazine ARs in livestock commodities were estimated using one-half tolerance residues in feed items and data from feeding studies. Processing factors for processed commodities were based on chemical-specific empirical data, with the exception of dried beef, for which the default was used. One hundred percent crop treated was assumed for all commodities. EPA's single-chemical dietary (food only) assessments are considered to be partially refined because the only departure from the most conservative default assumptions is the use of ARs instead of tolerances. For partially refined assessments such as these, the 95th percentile of the exposure distribution is the threshold for regulation (*8*).

DEEM contains modules to calculate acute (single-day) and chronic dietary exposures. DEEM is not able to estimate multi-day rolling average exposures. With DEEM, the best available approach to approximate a 4-day average available is to use the 2-day average food consumption data in the DEEM acute module. EPA then summed the 95th percentile exposures from the single-chemical dietary (food only) assessments for simazine and atrazine to approximate 95th percentile cumulative chlorotriazine dietary (food only) exposures. Table 1 summarizes EPA's "4-day" cumulative dietary (food only) exposure estimates for the chlorotriazines. Children 1-2 years old are the most exposed subpopulation, followed by infants (<1 year old).

Table 1. EPA Estimates of "4-Day" Cumulative Chlorotriazine Dietary (Food Only) Exposures. SOURCE: Adapted from Table 3.2 in Reference (*1*)

Population Subgroup	*DEEM 2-Day Average Acute Dietary (Food Only) Exposure (95th Percentile) (mg/kg/day)*		
	Atrazine (Single Chemical)	*Simazine (Single Chemical)*	*Cumulative (Sum of Atrazine and Simazine)*
All infants (<1 yr)	0.002176	0.001749	0.003925
Children 1-2 yr	0.003336	0.002536	0.005872
Children 6-12 yr	0.001638	0.001113	0.002751
Youth 13-19 yr	0.000918	0.000599	0.001517
Females 13-49 yr	0.000695	0.000542	0.001237

It is essential to understand that there is nothing wrong with EPA's 2018 single-chemical or cumulative dietary assessments. The partially refined single-chemical and cumulative dietary exposure estimates summarized in Table 1 minimized time and resource requirements while also enabling EPA to make safety findings and propose registration review decisions for the chlorotriazines. Nevertheless, the key simplifications in EPA's assessments (i.e., approximating 4-day average exposures using the 2-day average consumption data, summing 95th percentile exposures from two single-chemical assessments to approximate the 95th percentile cumulative exposure, and assuming that all corn commodities are treated with both atrazine and simazine) represent opportunities to incorporate modest refinements or improvements to the dietary assessments that can be achieved with CARES NG. Thus, this case study was conceived to illustrate and explore the capabilities of CARES NG that can yield more representative exposure estimates than can be achieved with DEEM.

CARES NG—Overview

CARES NG is a probabilistic, scientific, cloud-based software model that facilitates multi-source, multi-route aggregate (for individual chemicals) and cumulative (for multiple chemicals) exposure and risk assessments (9). The CARES NG Development Organization, a 501(c)(6) not for profit business entity formed for the specific purpose of developing and maintaining the CARES NG model, is closely collaborating with Creme Global (Dublin, Ireland) for the software/model development. Creme Global is a privately held, independent company specializing in the design and build of scientific and predictive models and data management systems for commercial and regulatory purposes. The CARES NG Development Organization is composed of the following member companies and task forces: BASF; Bayer Crop Science; the Council for the Advancement of Pyrethroid Human Risk Assessment, LLC (CAPHRA); Corteva Agriscience; the Outdoor Residential Exposure Task Force, LLC (ORETF); the Pyrethrin Joint Venture (PJV); the Pyrethroid Working Group (PWG); the Residential Exposure Joint Venture (REJV); Syngenta; and Valent. CARES NG has been developed through careful consideration of regulatory perspectives on the conduct of cumulative exposure assessments (*10*). The EPA has been kept informed during development of CARES NG to ensure that the model and resulting exposure assessments will be useful for regulatory purposes.

Table 2. Comparison of Commodities Included in the Atrazine and Simazine Single-Chemical Dietary Assessments, Including Tolerances and Anticipated Residues

Atrazine		*Crop Group*	*Simazine*	
Tol/AR[a] (ppm)	*Commodities*		*Commodities*	*Tol/ARa (ppm)*
		Citrus fruit	Crop Group 10-10	0.04/0.02
		Pome fruit	Crop Group 11-10	0.03/0.015
		Stone fruit	Crop Group 12-12	0.1/0.05
		Berries and small fruit	Blackberry, Blueberry, Grape, Loganberry, Raspberry	0.2/0.1
			Cranberry, Currant, Strawberry	0.25/0.125
0.2/0.1	Macadamia nut	Tree nuts	Crop Group 14-12	0.07/0.035
0.2/0.1	Corn (field, pop), grain; Corn (sweet), K+CWHR[b]	Grains	Corn (field, pop), grain; Corn (sweet), K+CWHR	0.2/0.1
0.2/0.1	Sorghum, grain			
0.1/0.05	Wheat, grain			
0.05/0.025	Guava	Tropical and subtropical fruit	Avocado, Olive	0.2/0.1
0.2/0.1	Sugarcane	"Other"		
0.04/0.0084	Meat[c]	Meat commodities	Meat[c]	0.03/—[d]
0.04/0.0090	Meat byproducts[c]		Meat byproducts[c]	0.03/—[d]
0.04/0.0032	Fat[c]			
0.04/0.0072	Kidney[c]			
0.04/0.0090	Liver[c]			
0.04/0.029	Milk	Milk	Milk	0.03/—[d]
—/0.018	Freshwater finfish	Fish		

[a] "Tol" = tolerance; "AR" = anticipated residue. [b] K+CWHR = kernels plus cobs with husks removed. [c] Applies to cattle, goat, horse, and sheep commodities. [d] EPA determined that anticipated simazine residues in livestock consuming treated feed items are below the threshold for inclusion in the simazine dietary assessment.

The first Cumulative Aggregate Risk Evaluation System, which was called simply CARES, was developed in 2001 as a population, calendar-based, probabilistic model for estimating aggregate and cumulative exposure and risk across multiple routes (food, water, and residential) for pesticides. The CARES NG Development Organization's current goal is to advance aggregate and cumulative exposure and risk modeling methods and capabilities for the benefit of the scientific, regulatory, and public interest communities. The model is undergoing significant upgrades from its prior incarnation to meet current and future needs, including movement to a cloud-based system for easier updates; creation of a streamlined and flexible model to accommodate low- to high-tier assessments; updates of consumption, residue monitoring, and behavioral databases; incorporation of EPA's current standard operating procedures (SOPs) for residential pesticide exposures (*11*); and enabling minute-by-minute exposure estimates for all routes.

As a probabilistic model, CARES NG can incorporate variability and uncertainty into the exposure calculations, which results in a distribution of exposures across the population. For dietary exposure assessments, variability is most frequently introduced through the use of the full distribution of pesticide residue data on treated commodities and the full distribution of food consumption data for the population. As a calendar-based model, a dietary assessment in CARES NG involves creation of food consumption diaries for all subjects in the dataset. There are two methods to create these diaries in CARES NG – one involves randomly repeating a given subject's diet over the year and the other makes use of food consumption data for similar subjects to more fully reflect variability of food consumption over time for any subject. Choosing one food consumption extrapolation method over the other introduces uncertainty into the results. As a result, a probabilistic assessment produces a distribution of exposures for the subject population, which enables the exposure assessor to estimate, for example, the exposure at a specific percentile in the population, e.g., the 95th, in which 95% of the population has an exposure below this value. Although conducting probabilistic assessments is more data- and labor-intensive, such models permit more refined and realistic estimates of exposure than default-based deterministic assessment. As such, probabilistic exposure and risk assessment is a very useful tool in regulatory decision making (*12*).

CARES NG is available in the public domain and can be accessed at https://caresng.org/. It will ultimately be supported by user fees to enable full system functionalities. CARES NG comprises two models: the dietary exposure model and the residential exposure model. At the time of publication, the probabilistic dietary model is fully functional and the residential model is under development. When complete, CARES NG will support exposure assessments to pesticide residues in food and water and residential sources.

Notes about This Case Study

DEEM is also available embedded in a calendar-based exposure assessment program called CALENDEX. Although CALENDEX has been used by EPA to conduct cumulative dietary assessments for some CMGs, EPA did not use CALENDEX for the chlorotriazines. Therefore, this case study uses EPA's DEEM assessments for the chlorotriazines as the point of comparison and does not consider the capabilities of CALENDEX.

Like DEEM, CARES NG uses consumption data expressed in terms of agricultural food commodities that were derived from the "What We Eat in America" (WWEIA) Survey conducted as part of the National Health and Nutrition Examination Survey (NHANES) conducted by the Department of Health and Human Services (DHHS). Foods reported "as consumed" in the WWEIA Survey (e.g., apple pie) are converted using recipe translation files to generate the consumption data in the Food Commodity Intake Database (FCID) for defined food commodities or raw agricultural commodities (RAC) (e.g., apples, sugar, and flour). Additional information contained in the FCID for each RAC includes supplementary information such as the cooked status (yes/no), cooking method (e.g., baked, boiled, etc.) and food form (e.g., fresh, frozen, etc.). EPA's chlorotriazine assessments were conducted with DEEM version 3.16, which incorporates food consumption data collected in the period 2003-2008, whereas the CARES NG software contains the consumption data collected in the period 2005-2010. Both DEEM and CARES NG use only the data for survey participants who recorded two days of food consumption. Food consumption patterns change over time, but the different consumption data sets used in DEEM and CARES NG are not expected to have a major impact on exposure estimates.

Before the dietary exposure assessments conducted for this case study are discussed, it is important to understand that this case study is not meant to be a "how to" manual for probabilistic, calendar-based exposure assessments in general or for CARES NG in particular. Rather, the aim of this case study is simply to explore and illustrate some of the capabilities in CARES NG. The specific mechanics of conducting the dietary assessments described later are beyond the scope of this case study. Readers interested in developing proficiency in CARES NG assessments are referred to the CARES NG website (caresng.org) and the Support Center within the model, which contains comprehensive documentation for the model and instructions on how to use it.

Table 3. Summary of PDP Data for Atrazine and Simazine

Dataset	*Notes*
Atrazine (All commodity samples collected 2010-2019)	• 74,167 samples of 83 different commodities analyzed • Detectable residues reported in 95 samples (29 in commodities with tolerances, 63 in commodities without tolerances, and 3 in commodities for which tolerances are not applicable) • Detected residues in commodities with tolerances ranged from 0.0014 to 0.027 ppm
Simazine (All commodity samples collected 2010-2019)	• 71,762 samples of 82 different commodities analyzed • Detectable residues reported in 5 samples (3 in commodities with tolerances and 2 in commodities without tolerances) • Detected residues in commodities with tolerances ranged from 0.0085 to 0.012 ppm
Sweet corn (Samples collected 2010, 2014, and 2015)	• Atrazine was not detected in 1,261 samples • Simazine was not detected in 1,262 samples • Maximum limits of detection (LOD) were 0.016 ppm for atrazine and 0.03 ppm for simazine
Corn grain (i.e., field corn) (Samples collected 2007 and 2008 – most recent data)	• Atrazine was not detected in 1,290 samples • Simazine was not detected in 1,230 samples • LODs were 0.004 ppm for atrazine and 0.003 for simazine
Milk (Samples collected 2011, 2016, and 2017)	• Neither atrazine nor simazine was detected in 2,162 milk samples • Maximum LODs were 0.0064 ppm for atrazine and 0.0019 ppm for simazine.

Methods and Results—Step by Step

The registered uses of atrazine and simazine, as shown in Table 2, have very little overlap. The notable exceptions are field, pop, and sweet corn. Atrazine is also used on several animal feed items, and EPA included anticipated atrazine residues in meat and milk commodities in the atrazine dietary assessment. Simazine is registered on several fruit, tree nut, and berry crops. Simazine is used on field, pop, and sweet corn, but EPA determined that anticipated simazine residues in livestock consuming treated animal feed items are below the threshold for inclusion in the simazine dietary assessment.

The outermost columns in Table 2 list the tolerances (Tol) for each chemical (*13, 14*) and the ARs that were assigned for each commodity in the dietary assessments. The tolerances for the chlorotriazines tend to be derived from the LOQs for the parent and select chlorinated metabolites. Because chlorotriazines and their metabolites are rarely quantified in treated commodities, EPA generally assigned one-half of the tolerance as a conservative estimate of the AR. For potential residues of atrazine in livestock, conservatively estimated values were assigned as ARs. Atrazine was detected in 11 of 1,479 catfish samples collected by the U.S. Department of Agriculture (USDA) Pesticide Data Program (PDP) over three years (2008-2010), and EPA assigned the maximum detected residue of 0.018 ppm to all freshwater finfish commodities. All dietary assessments for the chlorotriazines were conducted as "food only" assessments.

The residue assignments made by EPA—and that were carried over into the CARES NG assessments conducted for this case study—are quite conservative. Consider the information obtained from searches of the PDP program database (*15*) as summarized in Table 3. The low frequency of detection of atrazine and simazine residues (<<1% of all samples) and the low residue values (≤0.027 ppm) on the rare occasions when residues were detected demonstrate the conservatism of assuming ARs at one-half the tolerance and 100% crop treated in the chlorotriazine dietary assessments.

Step 1. CARES NG Multi-day Average Dietary Exposure Assessment

The first step in this case study is to illustrate the impact of calculating 4-day average exposures. To calculate rolling averages, CARES NG creates a full year of daily consumption diaries for each subject in the sample population. Creation of these diaries is achieved with one of two options. With the temporal repeat diet, the consumption diary for each survey subject is created by randomly alternating that subject's two days of consumption data over the model year. With the temporal match diet, a subject's consumption diary is created by randomly assigning that subject's own consumption data as well as consumption data for similar subjects over the model year. The survey sample of children (>1 year) and adult subjects are binned by gender, and children 1-19 years are also binned by age. Infants are binned by breastfeeding status in 3-month age bands. Characteristics such as race/ethnicity, socioeconomic status, and other factors (depending on the subpopulation) are used to identify similar subjects within each bin based on a statistical procedure of nearest neighbor distance and Gower's dissimilarity (*16*). The food consumption data for each of these similar subjects is then randomly assigned when constructing 365-day food consumption diaries for each survey subject. Approaches for constructing longitudinal consumption diaries, such as described in a Technical Manual for the original version of CARES (*17*), were evaluated by a Scientific Advisory Panel (SAP) convened by the EPA in 2004 (*18*). The temporal match diet results in more variability over time than the temporal repeat diet and is therefore expected to better reflect the range of consumption patterns over a year. The dietary assessments in this case study were therefore conducted with the temporal match diet algorithm. Rolling multi-day average exposures are calculated over the user-specified number of days from the 365 daily exposure estimates for each subject for the specified interval.

Infants under one year of age are a special case. In CARES NG, infants are grouped into three-month age bands, and breastfeeding status and consumption data are adjusted over the model year to match characteristics of the original NHANES survey population. A 4-month old in the NHANES survey will always be 4 months old in assessments conducted with DEEM and the acute model in CARES NG. With the temporal match diet algorithm in CARES NG, however, a 365-day diary of consumption events is created for that infant by extrapolating back to birth and forward to 12 months of age. The process involves changes in the assignment of breastfeeding status and associated consumption data, as well as body weight in each quarter over the year. The temporal match diet algorithm for infants may lead to seemingly surprising results at the highest exposure percentiles, as will be discussed later.

The acute module in DEEM estimates exposures using each of the two days of consumption data for each subject. In other words, each subject in the survey is represented by two exposure estimates. The average consumption data over the two survey days for each subject is used to approximate the 2-day average exposure in the DEEM acute module. CARES NG can calculate rolling average exposures for any multi-day period (up to a limit of 364) specified by the user. Figure 1 illustrates the acute/1-day, 2-day, 4-day, and 7-day average atrazine dietary exposures for children 1-2 years old and females 13-49 years old. Compared to the DEEM 2-day estimates, the CARES NG 4-day average exposures are 12% and 19% lower for these subpopulations, respectively. The capability of CARES NG to calculate 4-day average exposures results in more realistic estimates of potential exposures for the 4-day period of interest for chlorotriazines than can be achieved with DEEM.

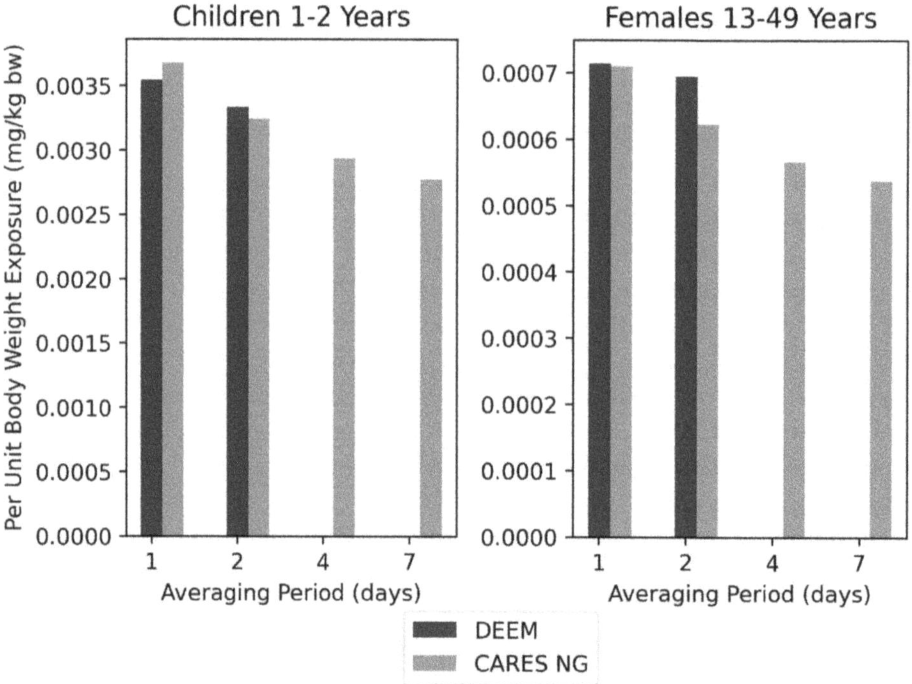

Figure 1. Impact of averaging period on multi-day dietary exposure estimates for atrazine.

Four-day average dietary exposure assessments were conducted in CARES NG for atrazine and simazine using the assumptions from the EPA assessments for ARs (listed in Table 2), processing factors (chemical-specific empirical factors for all processed foods except dried beef, for which the default was used), and percent crop treated (100% PCT). The upper portion of Table 4 shows EPA's dietary exposure estimates from the 2018 chlorotriazine cumulative assessment (*1*), and the lower portion shows the 4 day average exposures at the 95th percentile from CARES NG for each of the single chemicals as well as the sum. For all subpopulations except infants, the 4-day CARES NG estimates are reduced (11-21% for the single chemicals and 14-17% for the chlorotriazine sum) compared to the DEEM estimates. For infants, however, the CARES NG 4-day average exposures at upper exposure percentiles are higher than the corresponding DEEM 2-day exposures. This general result for infants is typically more noticeable when the number of averaging days is below 7 days (such as the 4 days in this case study). The causes of this pattern for infants are discussed further in the Discussion section.

Table 4. Comparison of DEEM and CARES NG Dietary (Food Only) Exposures for Single Chemicals and Chlorotriazine Sum

DEEM Dietary Exposure Estimates[a]			
Population Subgroup	*2-Day Average Acute Dietary (Food Only) Exposure (95th Percentile) (mg/kg/day)*		
	Atrazine (Single Chemical)	*Simazine (Single Chemical)*	*Sum of Atrazine and Simazine*
All infants (<1 yr)	0.002176	0.001749	0.003925
Children 1-2 yr	0.003336	0.002536	0.005872
Children 6-12 yr	0.001638	0.001113	0.002751
Youth 13-19 yr	0.000918	0.000599	0.001517
Females 13-49 yr	0.000695	0.000542	0.001237
CARES NG Dietary Exposure Estimates[b]			
Population Subgroup	*4-Day Average Dietary (Food Only) Exposure (95th Percentile) (mg/kg/day)*		
	Atrazine (Single Chemical)	*Simazine (Single Chemical)*	*Sum of Atrazine and Simazine*
All infants (<1 yr)	0.002534 (16%↑)	0.001946 (11%↑)	0.004480 (14%↑)
Children 1-2 yr	0.002937 (12%↓)	0.001994 (21%↓)	0.004931 (16%↓)
Children 6-12 yr	0.001463 (11%↓)	0.000907 (19%↓)	0.002370 (14%↓)
Youth 13-19 yr	0.000778 (15%↓)	0.000499 (17%↓)	0.001277 (16%↓)
Females 13-49 yr	0.000566 (19%↓)	0.000459 (15%↓)	0.001025 (17%↓)

[a] SOURCE: Adapted from Table 3.2 in Reference (*1*). [b] Comparison to corresponding DEEM estimate, in terms of percent decrease or increase, is shown in parentheses.

Step 2. CARES NG Cumulative Dietary Exposure Assessment—Assuming 100% Crop Treated for All Commodities

Table 5 shows the relative commodity contribution to 4-day average dietary exposures at the 95th percentile for children 1-2 years old, the most highly exposed subpopulation, from the two single-chemical assessments conducted in CARES NG. Corn commodities are the second highest contributor for each chemical, but there is otherwise no overlap of major contributors to exposure. The general lack of overlap in the atrazine and simazine registered uses diminishes the error introduced by approximating the cumulative assessment as the sum of the two single-chemical assessments. Nonetheless, conducting a cumulative assessment for atrazine and simazine with CARES NG represents a move toward more accurate exposure estimates.

Table 5. Commodity Contribution—Single-Chemical 4-Day Average Exposures (95th Percentile) Using CARES NG for Children 1-2 Years Old

Atrazine		*Simazine*	
Commodity	*Contribution*	*Commodity*	*Contribution*
Milk	69.8%	Berries	45.5%
Corn (field, pop, sweet)	16.2%	Corn (field, pop, sweet)	29.4%
Other grains	9.3%	Pome fruit	13.9%
Sugar cane	4.1%	Citrus fruit	7.6%
All others	0.5%	Stone fruit	3.1%
		All others	0.4%

The next step in this case study is to use CARES NG to calculate true cumulative dietary exposure estimates for the chlorotriazines. In the assessments discussed previously, the 95th percentile cumulative exposure estimate is approximated as the sum of the two single-chemical 95th percentile exposure estimates. The ability to set up a cumulative assessment within CARES NG is a distinct improvement over the approximation required with DEEM.

In creating the cumulative assessment, separate food residue files were created in CARES NG for each active ingredient. The single-chemical food residue files contain 49 commodities for atrazine and 104 commodities for simazine. For a cumulative assessment, however, CARES NG requires the same commodities in all food residue files. Because only corn commodities and macadamia nut may be treated with both atrazine and simazine, the list of commodities in the food files for the cumulative assessment contains 140 commodities [49 (atrazine) + 104 (simazine) – 12 (corn) – 1 (macadamia nut) = 140]. Zero residues are assigned to all pesticide-commodity combinations without permitted uses. ARs from Table 2 were assigned as point estimates in the food residue files. The assumption of 100% crop treated for all commodities was retained to be consistent with the EPA cumulative assessment. Although atrazine and simazine do not enjoy 100% market share on any registered commodity, the assumption of 100% crop treated justifies, in part, the selection of the 95th percentile exposures for regulatory purposes. Incorporating PCT estimates would trigger use of the 99.9th percentile exposure estimate for regulatory purposes (*8*), which would complicate comparisons for purposes of this case study.

Creation of a cumulative assessment in CARES NG requires the input of toxicology PODs for all chemicals. Table 6 lists the dietary (food) PODs for atrazine and simazine (*1*). These PODs were specified in the atrazine and simazine toxicology data table in CARES NG. CARES NG requires the selection of an index chemical, and atrazine was selected. When estimating cumulative exposures, CARES NG scales simazine residues to a toxicologically equivalent atrazine residue, and all reported cumulative exposures are in terms of atrazine equivalents. The dietary (food) PODs for atrazine and simazine are actually quite close, so the effect of this scaling is negligible for chlorotriazines. In fact, were the PODs not so close, EPA would have needed to introduce a scaling factor when summing single-chemical exposures to approximate the cumulative chlorotriazine exposures in terms of atrazine equivalents because the atrazine dietary (food) PODs were selected for risk characterization. Although this case study focuses only on estimated exposures, the ability of CARES NG to scale residues using toxicology PODs is a powerful feature.

Table 6. Dietary (Food) Points of Departure for Atrazine and Simazine

	Dietary (Food) Point of Departure (mg/kg/d)				
	Infants (<1 yr)	*Children 1-2 yr*	*Children 6-12 yr*	*Youth 13-19 yr*	*Females 13-49 yr*
Atrazine	3.06	3.24	2.57	2.33	2.29
Simazine	3.08	3.26	2.59	2.35	2.32

The 4-day average dietary exposures from the CARES NG chlorotriazine cumulative assessment, expressed in terms of atrazine equivalents, are shown in the rightmost column of Table 7. To the left, the DEEM 2-day average and CARES NG 4-day average exposure estimates from Table 3 are provided for comparison. With the exception of infants, CARES NG cumulative dietary exposures for chlorotriazines are reduced compared to EPA's DEEM estimates and the sum of the CARES NG single-chemical estimates. The CARES NG cumulative assessment shows reductions in 95th percentile exposures of 18-24% compared to the DEEM estimates for all subpopulations except infants, which show an 8% increase. The reasons for the results for infants will be discussed separately in the Discussion section.

Table 7. Comparison of the CARES NG Cumulative Assessment for Chlorotriazines to the Sum of the Single-Chemical Assessments Using DEEM and CARES NG (100% Crop Treated for All Commodities)

Population Subgroup	*Dietary (Food Only) Exposures (95th Percentile) (mg/kg/day)*		
	DEEM 2-Day: Sum of Atrazine and Simazine	*CARES NG 4-Day: Sum of Atrazine and Simazine*[a]	*CARES NG 4-Day: Chlorotriazine Cumulative*[a]
All infants (<1 yr)	0.003925	0.004480 (14%↑)	0.004225 (8%↑)
Children 1-2 yr	0.005872	0.004931 (16%↓)	0.004481 (24%↓)
Children 6-12 yr	0.002751	0.002370 (14%↓)	0.002258 (18%↓)
Youth 13-19 yr	0.001517	0.001277 (16%↓)	0.001219 (20%↓)
Females 13-49 yr	0.001237	0.001025 (17%↓)	0.000953 (23%↓)

[a] Comparison to corresponding DEEM estimate, in terms of percent decrease or increase, is shown in parentheses.

Step 3. CARES NG Cumulative Dietary Exposure Assessment—Percent Crop Treated Adjustment for Corn Commodities

The assumption of 100% crop treated for all commodities included in the previous assessment implies that all corn commodities are treated with both atrazine and simazine. EPA's average and maximum screening-level PCT estimates for field corn and sweet corn (6, 7) are shown in Table 8. These PCT estimates show that atrazine is used far more widely on field corn and sweet corn than simazine is. Given the much lower use of simazine, assumption of 100% crop treated for both active ingredients is unrealistic. The relative difference in the use of the two pesticides suggested the introduction of an adjustment that would preserve the 100% crop treated assumption for atrazine on corn while accounting for the lower use of simazine. Whereas EPA would typically use maximum estimates of PCT for an acute (short-term) assessment, the average PCT values were scaled to maximize the use of simazine relative to atrazine, as shown in Table 9.

Table 8. Percent Crop Treated Estimates for Field Corn and Sweet Corn

Chemical	*Commodity*	*Average PCT Estimate*	*Maximum PCT Estimate*
Atrazine[a]	Field corn	60%	70%
	Sweet corn	70%	75%
Simazine[b]	Field corn	5%	5%
	Sweet corn	2.5%	2.5%

[a] PCT estimates from EPA dietary assessment for atrazine (6). [b] PCT estimates from EPA dietary assessment for simazine (7).

Table 9. Calculation of Number of Samples to Include in Residue Distribution Table to Account for Relative Use of Atrazine and Simazine

Field Corn		*Sweet Corn*	
Chemical	*Average PCT*	*Chemical*	*Average PCT*
Atrazine	60%	Atrazine	70%
Simazine	5%	Simazine	2.5%
Sample Type	*Count*	*Sample Type*	*Count*
Samples needed	12	Samples needed	28
N with atrazine residues only	11	*N* with atrazine residues only	27
N with atrazine and simazine residues	1	*N* with atrazine and simazine residues	1

Using the scaling calculations in Table 9, a food residue distribution file for field corn was created containing 12 samples: 11 with the atrazine AR only and one with both atrazine and simazine ARs. There being no PCT estimates for popcorn, the field corn data were assigned as a surrogate. A food residue file for sweet corn was created containing 28 samples: 27 with the atrazine AR only and one with both atrazine and simazine ARs. Thus, 100% of corn samples are assumed to have atrazine residues, and a smaller fraction, in proportion to market share, is assumed to also have simazine residues. No corn sample has only simazine residues. Although the food residue files developed with this adjustment likely overstate the use of both chemicals on any corn crop, this adjustment retains the assumption of 100% crop treated in that all corn commodities in the assessment are assumed to be treated with one if not both chlorotriazines.

Other approaches are available to develop estimates of the co-occurrence of potential atrazine and simazine residues on corn commodities. Based on the average PCT estimates for field corn, assuming random use of these two pesticides, a 2×2 table suggests 57% treated with atrazine only, 2% treated with simazine only, 3% treated with both atrazine and simazine, and 38% not being treated with either. To preserve the 100% crop treated assumption for field corn, the untreated field corn is removed from consideration, and a food residue file could be created containing 57 samples with the atrazine AR, two samples with the simazine AR, and three samples with both atrazine and simazine ARs. A similar analysis for sweet corn would yield a food residue file containing 273 samples with the atrazine AR, three samples with the simazine AR, and seven samples with both atrazine and simazine ARs. In all likelihood, only a small fraction of corn might be treated with both atrazine and simazine, so one could also assume that no samples would contain both residues. Using this alternative approach and retaining the 100% crop treated assumption, a food residue file for field corn would have 12 samples with the atrazine AR, one with the simazine AR, and no samples with both ARs.

The simple scaling approach discussed previously corresponds to 8.3% and 3.6% of field corn and sweet corn samples in the food residue file, respectively, assumed to have been treated with both chemicals. The 2×2 analysis yields 4.8% and 2.5% of field corn and sweet corn samples, respectively, assumed to have been treated with both chemicals. The actual value is likely much closer to zero. The adjustment employed in this case study conservatively estimates the potential co-occurrence of residues and also preserves the 100% crop treated assumption that is necessary for the 95th percentile of exposures to be the appropriate threshold for regulation.

To create the cumulative assessment with the PCT adjustment for corn commodities, residue assignments for corn commodities in the two separate food residue files created for the prior assessments were modified. The point estimate ARs for all other commodities were retained. In place of the single point estimate AR for corn commodities in the atrazine and simazine food residue files, residue distribution tables were created with residue entries from the analysis in Table 9, as described previously. For each subject on each model day, CARES NG randomly selects a sample from the distribution table for each corn commodity and calculates the corresponding exposures with that value. This process is repeated for each iteration. The examples shown in this case study were run for five iterations. As discussed previously, the PCT adjustment applied to corn commodities is consistent with 100% crop treated because all corn commodities are assumed to be treated with atrazine, and some are also assumed to be treated with simazine. The preservation of the 100% crop treated assumption justifies, in part, the selection of the 95th percentile exposures for regulatory purposes.

Table 10. Comparison of CARES NG Cumulative Assessments for Chlorotriazines to the Sum of Single-Chemical Assessments with DEEM and CARES NG

Population Subgroup	*Dietary (Food Only) Exposures (95th Percentile) (mg/kg/day)*			
	DEEM 2-Day: Sum of Atrazine and Simazine	*CARES NG 4-Day: Sum of Atrazine and Simazine*[a]	*CARES NG 4-Day: Chlorotriazine Cumulative (All 100% PCT)*[a]	*CARES NG 4-Day: Chlorotriazine Cumulative (Adjusted PCT for Corn)*[a]
All infants (<1 yr)	0.003925	0.004480 (14%↑)	0.004225 (8%↑)	0.003661 (7%↓)
Children 1-2 yr	0.005872	0.004931 (16%↓)	0.004481 (24%↓)	0.004075 (31%↓)
Children 6-12 yr	0.002751	0.002370 (14%↓)	0.002258 (18%↓)	0.001877 (32%↓)
Youth 13-19 yr	0.001517	0.001277 (16%↓)	0.001219 (20%↓)	0.000955 (37%↓)
Females 13-49 yr	0.001237	0.001025 (17%↓)	0.000953 (23%↓)	0.000771 (38%↓)

[a] Comparison to corresponding DEEM estimate, in terms of percent decrease or increase, is shown in parentheses.

Table 10 summarizes the results of the CARES NG cumulative assessment, including the PCT adjustment for corn commodities in the rightmost column. The other columns present the exposure estimates from all prior steps in this case study for comparison. With the PCT adjustment for corn commodities, CARES NG yields cumulative dietary exposures that are reduced compared to EPA's DEEM approach for all subpopulations, including infants. Although the 7% reduction for infants is modest compared to the 31-38% reductions for the other subpopulations, this case study demonstrates an approach that may yield reduced exposures for all subpopulations.

Discussion

General Discussion

As stated earlier, the idea for this case study did not arise from an explicit need to refine EPA's existing single-chemical and cumulative chlorotriazine dietary assessments for the purposes of making a safety finding. The partially refined dietary assessments conducted by EPA using DEEM were sufficiently refined to support the interim registration review decisions for the individual chlorotriazines that were issued in September 2020 (*19–21*). Nevertheless, the approaches taken by EPA in their dietary assessments and the limitations of DEEM suggested an opportunity to explore and illustrate some of the capabilities of the CARES NG dietary model. With the exception of the final step in this case study, EPA's assumptions regarding ARs, processing factors, and 100% crop treated were preserved, and the case study results therefore illustrate primarily the effect of the capabilities of CARES NG. Because the version of DEEM used by EPA (version 3.16) and CARES NG contain food consumption data collected from slightly different but overlapping periods (2003-2008 for DEEM and 2005-2010 for CARES NG), the differences in consumption data sets will influence differences in exposure estimates between the two models, but the relative impact of the consumption database differences on exposure estimates is likely to be smaller than the changes arising from the use of the more advanced modeling capabilities within CARES NG.

The first step of the case study highlights the capability of CARES NG to calculate multi-day rolling average exposures. Whereas DEEM can only run the acute module with single-day or 2-day average consumption data, CARES NG can create 365-day consumption diaries for each subject in the NHANES survey. From the diaries, CARES NG then calculates daily exposures and from these can calculate rolling averages over any interval specified by the user at model runtime. The exposure distribution generated by the model can be thought of as a distribution of n-day average exposures, where n is the rolling average period specified by the user. The CARES NG assessments conducted for the case study showed that 4-day average exposures at the 95th percentile are lower than the corresponding 2-day DEEM estimates for all subpopulations except infants (discussed further in "The Special Case of Multi-Day Assessments for Infants"). The reductions compared to DEEM were modest, 11-21% for single-chemical assessments and 14-17% for the sum of the two chlorotriazine assessments, but there could come a time when a modest reduction in the dietary exposure is the difference between a failing and passing risk assessment. Use of the multi-day capability of CARES NG is a small first step to consider in situations such as that posed by the chlorotriazines, for which the averaging period is relatively small but is nonetheless long enough that the 2-day average approach in DEEM overstates potential exposures.

The next step in the case study highlights the capability of CARES NG to perform a cumulative assessment of exposures to multiple chemicals. EPA approximated the 95th percentile cumulative chlorotriazine exposures as the sum of the 95th percentile estimates from each of the two single-chemical assessments. Conducting a true cumulative assessment removes the uncertainty introduced by the simple summation of two distributions, as EPA did in the 2018 cumulative chlorotriazine assessment. Compared to EPA's assessment, the CARES NG 4-day average cumulative exposures at the 95th percentile are 18-24% lower for all subpopulations, with the exception of infants (discussed further in "The Special Case of Multi-Day Assessments for Infants"). The cumulative assessment capability of CARES NG thus represents another option to consider if a simple refinement of exposures is needed to yield a passing assessment.

The final step of the case study is the only one in which input assumptions from EPA's assessment were not entirely preserved. In this case, a modest adjustment was made to account for the fact that simazine is far less widely used on corn commodities than is atrazine. A simple approach was devised to preserve EPA's assumption of 100% of corn commodities being treated with atrazine while also accounting for the less frequent use of simazine. As a first conservative approximation, residue tables were created containing one sample with both atrazine and simazine ARs and the remainder with just atrazine ARs. When this PCT adjustment is made for field, pop, and sweet corn commodities, the resulting 4-day average cumulative exposures at the 95th percentile are 7% lower for infants and 31-38% lower for all other subpopulations than the corresponding DEEM estimates. Because the PCT adjustment applied to corn commodities is conceptually consistent with 100% crop treated (i.e., all corn commodities are assumed to be treated with atrazine, and some are also assumed to be treated with simazine), the 95th percentile exposures remain appropriate for regulatory purposes.

Special Case of Multi-day Assessments for Infants

As mentioned previously, some interesting exposure estimates were obtained for infants in the CARES NG 4-day rolling average exposure assessments conducted with the temporal match algorithm for food consumption data. It seems counterintuitive that the CARES NG 4-day average exposures for infants at upper percentiles would be higher than those from the DEEM 2-day average estimates. Understanding the reason for this surprising behavior requires a deeper dive into the mathematics and methodology used to create the 365-day consumption diaries in the infant population.

When the 365-day consumption diary is created for a subject in the selected population using the temporal match algorithm, there is a specific subset of the population from which each day of consumption data can be sampled. These subsets were created using specific criteria to ensure that the available sampling pool consists only of subjects who are sufficiently closely related. In the case of the infant population, the subjects are first binned by breastfeeding status and 3-month age band, and then criteria of race/ethnicity and socioeconomic status are used to find suitable matches. This process ensures that the resulting matching subset for each subject infant contains other infants who have relatively consistent consumption patterns. The distribution of body weights for each of the four infant age band subpopulations is shown in Figure 2. The body weights appear to be normally distributed, and the mean weight increases for each subsequent age band.

When the 365-day consumption diaries are created for each subject, the eating events that are sampled are not normalized to take the body weight of the matched subject into account. In other words, the absolute consumption amounts, rather than normalized amounts, are extrapolated across subjects. This creates an interesting anomaly for the infant subpopulation. Since the weights of infants can vary significantly, even within a quarter, it is possible in the 365-day diary that the consumption events assigned for a particular subject were sampled from another subject with a higher body weight and, more than likely, higher overall consumption amounts.

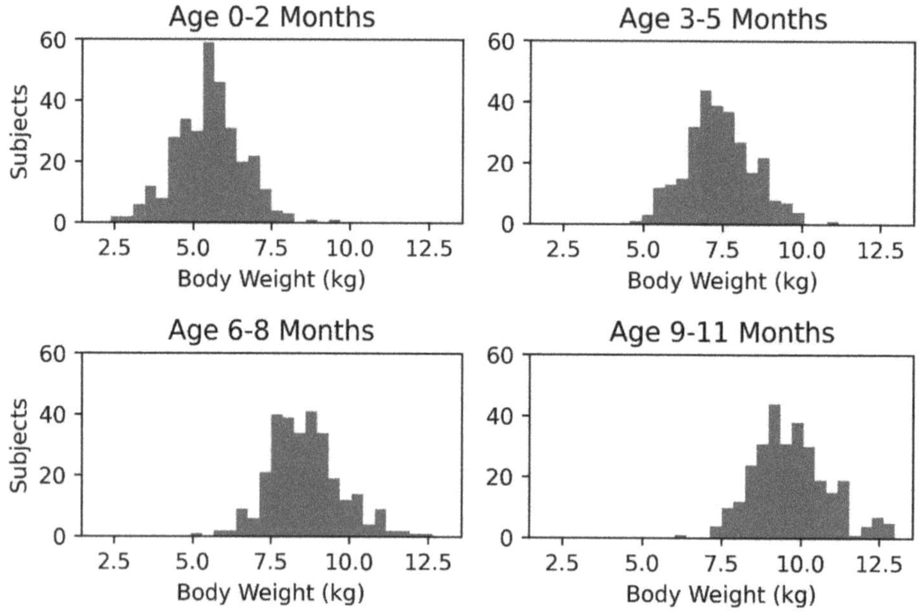

Figure 2. Distribution of infant body weights for each of the four age groups.

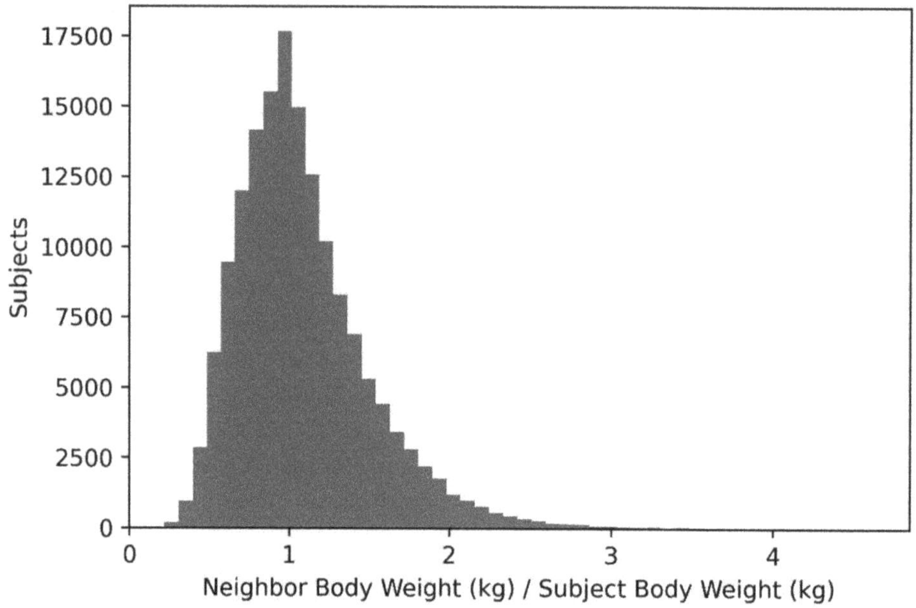

Figure 3. Distribution of neighbor body weights relative to subject body weights for the infant subpopulation.

To demonstrate this visually, the body weight for each subject neighbor was divided by the corresponding subject body weight to produce a distribution of ratios. Values greater than one (>1) indicate that the neighbor body weight is greater than the subject body weight, and values less than one (<1) indicate that the neighbor body weight is less than the subject body weight. As seen in Figure 3, the distribution of ratios for the infant subpopulation is skewed to the right, which indicates that, in the infant matching process, the body weights of the neighbors tended to be higher than the body weight of the subjects themselves. Heavier infants tend to eat more than lighter infants. When an absolute consumption value is assigned from a heavier neighbor and is then normalized by the subject's lower body weight, the result across the infant population is normalized consumption estimates that are higher at the upper percentiles of the distribution than is observed in the original NHANES survey population. Consequently, when the averaging period is short, such as the 4 days in this case study, the upper percentile exposure estimates from a CARES NG assessment may be greater than the corresponding acute and 2-day average estimates from a DEEM assessment.

Conclusions

CARES NG is the product of a collaboration between the CARES NG Development Organization and Creme Global. As such, CARES NG is an example of stakeholders coming together to develop models to advance predictive exposure assessments for regulatory purposes. The purpose of this case study is to illustrate some of the many capabilities currently available in the CARES NG dietary exposure assessment model. The exposure assessments conducted for this case study required only small changes in methods compared to EPA's original partially refined dietary (food only) assessments, but the exposure estimates were nonetheless reduced compared to the DEEM estimates, with a few exceptions for infants. Exposure assessors may wish to consider the method by which CARES NG conducts multi-day assessments for infants, particularly when the number of days in the averaging period is small (<7 days), to determine the most appropriate path forward for any particular assessment including infants (i.e., temporal repeat diet or temporal match diet). The CARES NG multi-day exposure estimates for infants are robust, but the assessor must be aware of how the method chosen to generate food consumption diaries impacts the manner in which exposures are estimated.

Although this case study focuses exclusively on exposure estimates, it is important to consider that the potential risks associated with the estimated exposures are of primary concern in a regulatory setting. The dietary model in CARES NG provides several potentially powerful options, and the approaches discussed in the assessments conducted for this case study suggest a few initial steps to be considered should dietary exposures require refinement to support a safety finding.

Acknowledgments

The authors would like to acknowledge technical input and support from Cronan McNamara, John O'Brien, and Clare Thorp of Creme Global. The authors would like to acknowledge the encouragement and support of Lynn Bergeson and Lisa Campbell of Bergeson & Campbell, P.C. and Kent Rupprecht of Bayer Crop Science and the CARES NG Development Organization.

References

1. Rickard, K.; Drew, D.; Liccione, J.; Donovan, W. Chlorotriazines: Cumulative Risk Assessment – Atrazine, Propazine, and Simazine. DP Barcode D447476. Office of Pesticide Programs, U.S. Environmental Protection Agency, Washington, DC, July 10, 2018.
2. Dietary Exposure Evaluation Model-Food Commodity Intake Database (DEEM-FCID), v. 3.18. https://www.epa.gov/pesticide-science-and-assessing-pesticide-risks/deem-fcidcalendex-software-installer (last accessed March 23, 2021).
3. Cumulative and Aggregate Risk Evaluation System–Next Generation (CARES NG). http://caresng.org/ (last accessed April 7, 2021).

4. *Guiding Principles for Performing Aggregate Exposure and Risk Assessments*; Office of Pesticide Programs, U.S. Environmental Protection Agency, Washington, DC, November 28, 2001.
5. *Guidance on Cumulative Risk Assessment of Pesticide Chemicals That Have a Common Mechanism of Toxicity*; Office of Pesticide Programs, U.S. Environmental Protection Agency, Washington, DC, January 14, 2002.
6. Donovan, W. Atrazine: Acute, 4-Day, Background, and Chronic Dietary (Food Only) Exposure Assessments for Registration Review. DP Barcode D442826. Office of Pesticide Programs, U.S. Environmental Protection Agency, Washington, DC, June 12, 2018.
7. Donovan, W. Simazine: Acute, 4-Day, Background, and Chronic Dietary (Food Only) Exposure and Risk Assessments for Registration Review. DP Barcode 442822. Office of Pesticide Programs, U.S. Environmental Protection Agency, Washington, DC, June 12, 2018.
8. *Choosing a Percentile of Acute Dietary Exposure as a Threshold of Regulatory Concern*; Office of Pesticide Programs, U.S. Environmental Protection Agency, Washington, DC, 2000.
9. CARES NG About Page. https://caresng.org/about/ (last accessed February 20, 2021).
10. *Framework for Cumulative Risk Assessment*, EPA/600/P-02/001F; Office of Research and Development, Center for Public Health and Environmental Assessment (CPHEA), formerly known as the National Center for Environmental Assessment (NCEA), U.S. Environmental Protection Agency, Washington, DC, 2003.
11. *Standard Operating Procedures for Residential Pesticide Exposure Assessment*; Office of Pesticide Programs, U.S. Environmental Protection Agency, Washington, DC, October 2012.
12. *Probabilistic Risk Assessment to Inform Decision Making: Frequently Asked Questions*; Risk Assessment Forum, Office of the Science Advisor, U.S. Environmental Protection Agency, Washington, DC, July 2014.
13. 40 CFR § 180.220. Atrazine; Tolerances for Residues. https://www.law.cornell.edu/cfr/text/40/180.220 (last accessed March 23, 2021).
14. 40 CFR § 180. 180.213. Simazine; Tolerances for Residues. https://www.law.cornell.edu/cfr/text/40/180.213 (last accessed March 23, 2021).
15. PDP Databases and Annual Summaries. https://www.ams.usda.gov/datasets/pdp/pdpdata (accessed February 2, 2021).
16. Gower, J.; Legendre, P. Metric and Euclidean Properties of Dissimilarity Coefficients. *J. Classification* **1986**, 3, 5–48.
17. *CropLife America. CARES (Cumulative and Aggregate Risk Evaluation System) – Technical Manual.* March 20, 2002. https://archive.epa.gov/scipoly/sap/meetings/web/pdf/tmanual2.pdf (last accessed March 23, 2021).
18. U.S. Environmental Protection Agency Scientific Advisory Panel – April 29–30, 2004: A Model Comparison: Dietary and Aggregate Exposure in Calendex, CARES, and Lifeline. https://archive.epa.gov/scipoly/sap/meetings/web/html/042904_mtg.html (last accessed March 23, 2021).
19. *Atrazine - Interim Registration Review Decision - Case Number 0062*; Office of Pesticide Programs, U.S. Environmental Protection Agency, Washington, DC, September 14, 2020.
20. *Simazine - Interim Registration Review Decision - Case Number 0070*; Office of Pesticide Programs, U.S. Environmental Protection Agency, Washington, DC, September 10, 2020.
21. *Propazine - Interim Registration Review Decision - Case Number 0230*; Office of Pesticide Programs, U.S. Environmental Protection Agency, Washington, DC, September 2, 2020.

Chapter 10

Agricultural Handler Exposure Task Force: A Technical Overview of Agricultural Handler Exposure Data Development

Eric D. Bruce*

Agricultural Handler Exposure Task Force Consultant
21 Oak Knoll Court, Walnut Creek, California 94596, United States

***Email: eybruce@pacbell.net**

The Agricultural Handler Exposure Task Force (AHETF) was established in 2001 by several pesticide manufacturers to generate agricultural pesticide handler exposure data meeting Environmental Protection Agency (EPA) data requirements for registration. Details on the formation and operation of the AHETF are described in a separate chapter authored by Barnekow. This chapter summarizes the technical aspects of AHETF's two-decades-long project of assembling a database of worker exposure data.

The scientific question addressed by AHETF was:

> "What is the expected magnitude and (statistical) distribution of worker exposure to pesticides during distinct occupational pesticide handling scenarios?"

AHETF worked closely with regulatory agencies from the United States, Canada, and California (collectively known as the Joint Regulatory Committee, JRC) to assemble worker exposure data that met current scientific quality criteria. These exposure data reflect dermal and inhalation exposure potential for about 25 distinct pesticide handler scenarios and were assembled into a searchable database called AHED, Agricultural Handler Exposure Database. The data within AHED are a combination of data from existing high-quality worker exposure studies and new studies designed and conducted by AHETF. Collectively, they provide sufficient measurements to define exposure potential for each handler situation tested. Because passive dosimetry techniques and non-volatile pesticide chemicals were used to measure exposure potential, the resulting data are considered generic and can be utilized by AHETF members and regulatory agencies to support exposure and risk assessments for a wide variety of pesticide products.

There is a glossary of terms and abbreviations near the end of this chapter.

Background

Occupational handler exposure data for use in calculating exposure and risk assessments have been an Environmental Protection Agency (EPA) Guideline requirement since 1987. At that time, the only way registrants could satisfy this requirement was to conduct product-specific exposure studies on workers who mix, load and apply pesticides. Based on the recognition that handler exposure to pesticides is generic rather than product-specific, the industry and the regulatory agencies cooperated in establishing the Pesticide Handlers Exposure Database (PHED) (*1*) in the early 1990s. Companies contributed exposure data without any proprietary or compensation protections.

By 2001, EPA, Canada's Pest Management Regulatory Agency (PMRA) and the California Department of Pesticide Regulation (CDPR) recognized there were serious limitations to PHED. The technical limitations of PHED are outlined as follows:

- Inadequate number of measurements for one or more body areas
- Inadequate quality assurance or quality control data (e.g., poor or missing field fortification data)
- Use of patch dosimeters instead of whole-body dosimeters
- Lack of entire body dermal estimates for workers (i.e., not all body parts monitored for dermal exposure for some subjects)
- Many non-quantifiable residues on dosimeters
- Lack of diversity for study conditions (e.g., same workers used repeatedly or all workers handling the same amount of product)
- Lack of representativeness of study conditions (e.g., products or procedures that are no longer in common use)

Therefore, in early 2001, the EPA announced they would be producing monographs of handler exposure scenarios, compiled from all available sources, to ultimately replace PHED, recognizing that the available data were inadequate (2).

The Agricultural Handler Exposure Task Force (AHETF) then assembled a database of occupational handler exposure measurements for a variety of distinctly different pesticide handling scenarios including a decade-long testing program. Much of the data reflect new measurements collected in studies conducted by AHETF, but some of the data are from existing studies that met stringent quality and usefulness criteria. Most of the data are assembled in a database tool called AHED® (Agricultural Handler Exposure Database). The only AHETF data not included in AHED pertain to seed treatment procedures, such as bagging treated seed.

The scope of the AHETF testing program, that is determining exactly which handler scenarios would be covered in their database, was determined from task force member surveys about what types of data are needed to support their products along with discussions with the Joint Regulatory Committee (JRC, a committee including task force members and interested regulatory agencies) about which scenarios needed updated exposure data. Testing priorities were also jointly developed by AHETF members and the JRC.

The purpose of this chapter is to provide a summary of important technical issues and decisions that guided the data development by AHETF.

Human Studies Rule

EPA's Human Studies Rule is the common term for amendments to regulations involving intentional exposure to pesticides at 40 CFR (Code of Federal Regulations) 26 that became effective April 7, 2006 (*3, 4*). The Human Studies Rule was revised in 2013; the revisions became effective April 15, 2013. This regulation added new requirements for studies with human subjects, including obtaining informed consent from subjects, requiring review of proposed research by an institutional review board, and prohibiting EPA from relying on research involving intentional exposure of human subjects unless the research was ethically and scientifically valid. The Human Studies Rule also included the establishment of a Human Studies Review Board (HSRB) to review study protocols and final reports.

The regulation (40 CFR §26.1102(i)) defines "research involving intentional exposure of a human subject" as: "a study of a substance in which the exposure to the substance experienced by a human subject participating in the study would not have occurred but for the human subject's participation in the study." EPA determined that if AHETF asks a worker to do anything they were not going to do anyway (except allow the monitoring techniques to be used by researchers), then the study involves intentional exposure. But, if the subject would be exposed to the pesticide in exactly the same way as if he did not participate in the study, then the study was considered to be "observational" and not subject to HSRB review. Since AHETF often controls one or more aspects of the handling task (such as product used, amount of product handled, equipment used, or date of handling the product), AHETF studies are often subject to the new regulation requiring HSRB review. AHETF studies that are not observational are commonly referred to as "scripted" studies.

The HSRB is chartered under the Federal Advisory Committee Act to provide advice, information and recommendations to EPA on issues of research involving human subjects, including: research proposals and protocols, reports of completed research with human subjects, and how to strengthen EPA's programs for protection of human subjects of research. Under the Human Studies Rule, HSRB is required to review all protocols for third-party research involving intentional exposure of human subjects to a pesticide when that party has the intention of submitting the resulting data to EPA.

Members of the Board are selected by EPA and are nationally recognized experts in relevant scientific or technical disciplines such as biostatistics; human toxicology, including pharmacokinetic and toxicokinetic studies, clinical trials, and toxicology of cholinesterase inhibitors and other classes of environmental substances; bioethics; and human health risk assessment. In selecting HSRB members, EPA considers experts from the environmental scientific/technical fields, human health care professionals, academia, industry, public and private research institutes or organizations, other governmental agencies, and other relevant interest areas.

Soon after its formation, the HSRB recommended significant changes in the overall AHETF testing program and study procedures in use at the time. Those changes resulted in several important differences between how the pre-rule and post-rule exposure data were collected. The most important changes involved: evaluating risks to participants, risk mitigation procedures, recruitment procedures, equitable selection of subjects, sample size determinations, and statistical analysis procedures. Because of the important impact of the Human Studies Rule on the AHETF testing program, discussions throughout this chapter often compare "pre-rule" and "post-rule" practices.

The HSRB reviewed AHETF research (protocols and completed studies) as required under the Human Studies Rule, and after each meeting produced a final report that summarized the advice and recommendations for each topic reviewed. The complete reports from the HSRB are available at epa.gov (search for HSRB and "final report" and the date of the meeting). The HSRB also prepared and distributed minutes of each meeting, but the meeting reports reflected final decisions.

Post-rule, AHETF conducted two studies that were "observational" in nature, which meant they fell outside the official purview of HSRB. These studies were an open cab groundboom application of liquids with soil incorporation study (designed to replace five invalid monitoring units, or MUs, from an earlier study) and a closed cockpit aerial application of liquids study. For these two studies, EPA confirmed in advance that AHETF would not ask workers to alter their planned work routines in any way, except to allow researchers to collect passive dosimetry samples. Materials for observational studies were sent to EPA for both science and ethics review, exactly as for intentional dosing studies; however, the HSRB did not review the protocols or final reports for either of these studies.

Existing Study Review and Purchase

Prior to conducting any field studies with human volunteers, AHETF reviewed existing handler exposure studies from various sources (primarily from AHETF members, CDPR, and the open literature) and acquired the rights to cite data that met established acceptance criteria. In addition, data from PHED studies were always reviewed to determine whether those data were suitable for the AHETF generic database. This section discusses the process by which AHETF, in cooperation with the Joint Regulatory Committee (JRC), evaluated existing data to determine their acceptability in a generic database.

These data reviews were very detailed and involved a consensus process among the JRC to ensure consistent criteria were applied to all existing studies. Although some useful worker exposure studies were acquired by AHETF, most of the existing data were not sufficient to meet the generic data needs identified in advance by the JRC. Over 200 worker exposure studies were reviewed against established acceptance criteria: five studies provided 82 MUs for a variety of non-seed treatment scenarios and another 15 studies provided 466 seed-treatment-related MUs.

Review Process—Non-seed Treatment Studies

The process for evaluating existing studies which might provide exposure data for a proposed AHETF scenario involved the following four steps:

Step 1: Development of Data Acceptability Criteria

During 2001, AHETF developed formal acceptance criteria for determining whether existing exposure studies include data that are suitable for a generic database. In summary, the acceptance criteria required all of the following:

- Study conducted under Good Laboratory Practices (GLPs), or with similar technical controls
- Tasks, equipment, and formulations that are well-described and representative of North American practices and conducted by experienced handlers
- Supporting information for monitoring details such as times, amount of product handled, worker observations, weather conditions, etc.
- Dosimetry that provides whole-body exposure estimates, including for hands and head
- Field fortification data for each exposure matrix with acceptable recoveries and for triplicates at two levels, or more
- Monitoring times that reflect half a typical work period

- Quantifiable residues in most exposure samples, or sufficiently low limits of quantitation
- Validated analytical methods and laboratory fortifications that indicate the methods are working properly
- Limited use of personal protective equipment (PPE): evaluated on a scenario-by-scenario basis

AHETF and the JRC also determined there was no way to utilize biomonitoring-only studies in a generic database. Therefore, none of those studies were reviewed in detail nor accepted for citation by AHETF, and therefore the AHETF data sets contain only passive dosimetry results.

Step 2: Primary Review

The primary review was conducted by the study submitter (or a designated representative) and simply reviewed the study against the acceptance criteria outlined earlier. If the study passed the primary review, then a complete copy of the study report was provided to AHETF for consideration. The purpose of the primary review was to avoid the submission of studies that clearly did not meet the selection criteria. It also prompted study submitters to check on the availability of supporting information such as raw data records.

Step 3: Secondary Review

The secondary review was conducted by an AHETF contractor hired and trained for this purpose, and was conducted under a confidentiality agreement with study owners. Consultants with experience conducting worker exposure studies or with experience as analytical monitors were employed for these technical study reviews. To maintain the confidentiality of proprietary reports during the review process, AHETF assigned a code to each study and the AHETF Manager maintained a file that linked the assigned AHETF study numbers to the identity of the original study report (including its title, report number, owner, etc.).

The purpose of the secondary review was to verify the accuracy of the primary review and, where necessary, provide a more detailed discussion of each specific part of the acceptance criteria, including whether each criterion was met and possible deficiencies in the study data. The secondary review was done in two stages. First, a Secondary Review Form (SRF) was generated. This was a brief document in a standard format that summarized the study in terms of the scenario(s) involved, equipment used, formulations tested, dosimetry used, clothing/PPE worn, technical issues to discuss, and recommendations about possible acceptability. Second, a more detailed Secondary Review Report (SRR) was prepared for those studies that were possibly acceptable based on the initial SRF.

Step 4: Final Review

The final review and decision on whether a study was accepted for purchase was made by the JRC. The SRRs prepared by AHETF contractors and, when available, regulatory reviews of contributed studies (e.g., from EPA or PMRA when the study had previously been submitted to a regulatory agency) were provided to the JRC to serve as the basis for the final review. Reviews considered study-wide technical issues in addition to MU-specific issues and sometimes resulted in only some MUs within a study being acceptable. Decisions were made at a few JRC meetings and on many special conference calls, mostly during the time period of November 2003 to June 2004. Decisions were documented in both SRFs and SRRs. During this period, the studies evaluated involved a variety of scenarios within the AHETF scope, except that seed treatment study reviews were conducted later.

Review Process—Seed Treatment Studies

Seed treatment studies were evaluated using a similar process during the period of April 2012 to September 2013. However, two important changes in the review process were made. First, AHETF identified a few general issues that impacted multiple existing seed treatment studies and warranted discussion in general terms prior to completing the secondary reviews of the seed treatment studies. These issues were discussed by the JRC and included special issues such as whether full-face protection could be acceptable, how to handle a large number of worker samples with non-quantifiable residues, and how much field fortification data are necessary for these indoor studies. For most of the issues discussed on the call, it was difficult to make a hard rule that would apply to all studies, so JRC agreed to make final decisions on a case-by-case basis and all at once after all seed treatment studies had been reviewed.

The JRC then held a few meetings to discuss proposals for seed treatment scenario definitions and a variety of specific technical acceptance issues for seed treatment studies and/or MUs. In addition, a series of conference calls took place to discuss study-specific acceptability issues and to review and approve existing study SRFs and SRRs.

Documentation of Final Reviews

Documentation of all existing study reviews, discussions, and decisions (e.g., SRFs, SRRs, meeting minutes, and conference call notes) was maintained by AHETF. SRRs for all studies that were approved for purchase by AHETF were submitted to the EPA and so they have Master Record Identification Numbers (MRIDs) assigned to them by EPA. SRRs were also referenced and linked to scenario monographs that rely on those data. The SRRs and monographs were also submitted to PMRA and CDPR. In addition, sponsor study reports for studies that were purchased by AHETF were submitted to EPA (and so have MRIDs assigned) – sometimes this was a submission made by the original study sponsor, but sometimes (e.g., for several seed treatment studies) the submission of the sponsor report was made by AHETF (because the study sponsor had never submitted a report to EPA).

Exposure Monitoring Techniques

When AHETF began collecting MUs for its testing program in 2003, it utilized standard passive dosimetry techniques that previous generic exposure task forces had used, including: inner whole-body cloth dosimeters, hand washes, face/neck wipes, and personal air sampling devices. Post-rule, AHETF monitoring was temporarily suspended while EPA reviewed AHETF study designs and monitoring techniques to ensure they were scientifically and ethically valid. An important scientific aspect was a re-evaluation of the specific monitoring techniques being utilized, and EPA concluded the passive dosimetry techniques were indeed valid.

The following sections describe the results of EPA's evaluation of AHETF's monitoring techniques and describes in detail the techniques used by AHETF. It also discusses the specific attributes of acceptable surrogate chemicals that make them useful within a generic database of exposure data.

Validity of Passive Dosimetry

Human exposure to pesticides is typically measured using passive dermal and inhalation dosimetry techniques and involves a specific pesticide exposure situation experienced by a pesticide user. Passive dosimetry typically consists of a partial or whole-body dosimeter (usually undergarments, but sometimes dermal patches) on the skin under clothing, washes of the hands, swabbing of the face and neck with moistened gauze pads or absorbent hat patches, and collecting air samples in the breathing zone of the pesticide user with a personal air sampling pump and sampling device (such as a filter and/or chemical sorbent). There was concern among some interested individuals, including the HSRB, that passive dosimetry may not have been valid for its intended purpose, which is to enable a researcher to estimate chemical exposure and absorbed dose in safety assessments. In response to these concerns, AHETF and the Outdoor Residential Exposure Task Force (ORETF) compared the absorbed doses calculated from passive dosimetry (PD) with urinary metabolite biomonitoring (BM) for several studies.

Concurrent, parallel, or sequential passive dosimetry was conducted in addition to the biomonitoring data for the workers in the studies. This allowed for a direct comparison between measured absorbed dose and an absorbed dose estimated from passive dosimetry (after application of dermal absorption and metabolism/pharmacokinetic assumptions).

The task forces concluded, as confirmed by comparison with biomonitoring, that passive dosimetry estimates of dermal and inhalation exposure, and subsequent absorbed dose, were valid and useful for the intended purpose in the current regulatory paradigm. As a result, there was no need to develop validation protocols or to adjust passive dosimetry estimates.

ORETF prepared a comparison for biomonitoring results from 24 scenarios against passive dosimetry estimates from PHED for the same scenario. These comparisons were done on a scenario-wide basis and a ratio of PD/BM estimates was calculated for each scenario. Overall, it was concluded that exposure estimates derived from generic passive dosimetry studies provide a reasonable estimate of exposure, as confirmed by comparison to biomonitoring data for analogous scenarios.

Reports of the comparisons summarized earlier were discussed at a Scientific Advisory Panel (SAP) meeting in January 2007, at the request of EPA (5). The final minutes from this SAP meeting (6) indicated the following:

> "The Panel concluded that although a bias may exist, no bias between dermal exposure monitoring and biological monitoring could be detected in large part because of the statistical uncertainty inherent in the exposure data."
>
> "The relationship between passive dosimetry and biomonitoring depends in part upon what form of dosimetry is being used and upon the question being asked. Within this caveat, the Panel concluded that passive dosimetry can generate data that can be used to predict worker exposure for a wide variety of scenarios and activities."

The SAP minutes also indicated the following with regards to specific monitoring techniques:

> "The Panel was slightly more equivocal about a need to correct handwashing for its efficiency at recovering pesticides from the skin."
>
> "Overall, the Panel recommended that use of a hand washing technique should be accepted in AHETF or AEATF studies if it is supported by either laboratory data and/or a model that predicts and can correct for its efficiency over the sampling time for the pesticide being

studied." (Note that AEATF refers to another pesticide industry worker exposure task force called the Antimicrobial Exposure Assessment Task Force that also uses hand washes.)
"The Panel concluded that generally there is no need to correct whole-body dosimeters worn under outer garments for the breakthrough of residues." "The use of a whole-body dosimeter placed directly against the skin is recommended, but no sure means of detecting its breakthrough from liquid saturation was identified."
"The Panel also concluded that biological monitoring can be complementary to dermal exposure monitoring to detect if not estimate the amount of breakthrough from a whole-body dosimeter (WBD). Therefore, the Panel generally supports an Agency proposal that biomonitoring may be included in any sampling plan, but recommended that it not be required because of its potential to bias participant selection."

Based on the SAP discussions and final recommendations shown earlier, AHETF continued to utilize the passive dosimetry techniques it was using prior to the passage of the Human Studies Rule (i.e., pre-rule). AHETF believes all the passive dosimetry techniques it used are acceptable and do not need any adjustments for method efficiency or breakthrough.

Passive Dosimetry Techniques

The following sections describe the various exposure monitoring techniques used for the studies that were conducted or purchased by AHETF. All of the studies used some form of passive dosimetry to collect exposure residues from the worker's body, head, and hands. In addition, some studies also measured exposure to the feet. All studies also used personal air sampling pumps and filters and/or sorbents to trap residues from air in order to estimate inhalation exposure.

Whole-Body Exposure

All of the AHETF-conducted studies involved the use of one-piece, all cotton, inner whole-body dosimeters (WBDs). The WBDs were worn under clothing which was provided by the subject (except in a few cases) and were used to assess potential exposure to the worker's skin. The WBDs were cut into either two or six sections for independent chemical analysis, depending on the expected magnitude of dermal exposure.

In the studies that were purchased by AHETF, inner WBDs were also utilized. However, they were sometimes two-piece garments and the sectioning/analysis varied somewhat with one, three, four, or five pieces analyzed. In all cases, however, the entire body from ankles to wrists to neckline was represented by an inner WBD.

Foot Exposure

For a few scenarios, AHETF employed sock dosimeters (new cotton socks provided by AHETF) to estimate exposure to the feet. Exposure to this body area is generally insignificant for pesticide handlers since shoes and socks (required by the Worker Protection Standard, WPS) provide considerable protection. However, scenarios that involved handheld application equipment and workers who walked while they sprayed (such as backpack applications) included this dosimetry technique. Cotton socks were worn under normal footwear (i.e., shoes/boots and socks) and were collected at the end of the day for analysis, similar to the WBDs.

Hand Exposure

To estimate hand exposure, AHETF selected a standard hand washing technique using a dilute surfactant solution. First, the worker was asked to rub their hands together while a researcher slowly poured 400 mL of diluted Aerosol OT solution (AOT) in water over his hands. The wash solution was collected in a bowl placed under the worker's hands. Researchers were trained to take about 30 seconds to pour the liquid. Next, the worker was asked to immerse and rub his hands in the wash water in the bowl for about 30 seconds. Finally, 100 mL of fresh surfactant solution was poured over the worker's hands to rinse them, and the rinse water was collected in the same bowl. The combined wash/rinse water was transferred to a sample container and analyzed to determine total residues. Pre-rule studies did not require a hand wash sample 'before smoking'. However, based on the HSRB's recommendations following the October 2010 meeting (7), AHETF added the requirement that a hand wash sample be collected before a subject smokes.

In the studies purchased by AHETF, the hand wash techniques varied somewhat and sometimes included different types of solvent systems (e.g., ethanol or diluted Esemtan®), different volumes of solution (e.g., up to 1,000 mL total), or the use of plastic bags containing a wash solution and in which workers rubbed their hands instead of the pouring/rubbing technique. EPA and JRC agreed these techniques were similar enough to current standard techniques and were acceptable.

AHETF reviewed human in vitro and rat in vivo and in vitro dermal absorption studies and concluded the skin wash off procedure in the human in vitro studies removed a mean of 91.2% of the applied dose after an 8-hour exposure period. The wash off procedures used in dermal absorption studies are analogous to the procedures used in AHETF worker exposure monitoring studies. Therefore, the hand wash and face/neck wipe procedures used by AHETF are expected to achieve near complete recovery of residues from the surface of the skin and residue values do not require adjustment.

Head Exposure

AHETF selected a standard face/neck wiping technique involving two successive wipes with cotton gauze pads wetted with 4 mL of a dilute surfactant solution (same solution as hand washes: diluted AOT in water). The entire face, front of the neck, and back of the neck were wiped two separate times and the two gauze pads were combined for analysis. This technique was preferred to using hat patches since no extrapolation of residues to the entire face/neck area was needed.

For a few AHETF studies, the study design involved the use of chemical-resistant (CR) headgear. This headgear was a requirement for all open cab airblast study MUs, but for the greenhouse/nursery application scenario, CR headgear was required only for MUs involving overhead application. When CR headgear was worn, AHETF utilized both inner patches (under the hat) and outer patches (on top of the hat) that were arranged so one was not right on top of the other. Residues found on these patches provided a measure of inner and outer potential exposure for the part of the head not accounted for by the face/neck wipe. This allowed estimation of total head exposure to workers who might or might not wear chemical-resistant headgear. The latter required summing extrapolated exposure from the inner patch with that from the outer patch. Cut-up pieces of a whole-body inner dosimeter were used as head patches.

Most of the studies purchased by AHETF involved face/neck wiping procedures; however, the techniques varied somewhat and sometimes included different types of wiping material (e.g., a cotton loop towel), different solvent systems (e.g., Esemtan® or liquid hand soap), or different volumes of solution on the media (e.g., up to 20 mL). Two purchased studies involved only head patches to determine head exposure. In these cases, patch residues were extrapolated to the entire head and neck.

Inhalation Exposure

AHETF selected two standard inhalation dosimetry techniques that were used extensively in previous exposure monitoring activities. Each technique involved a personal air sampling pump and a collection device attached to the worker's collar and facing downward. Air was drawn through the device at 2 liters per minute (L/min).

The collection device used for almost all surrogates was a glass OSHA Versatile Sampler (OVS) tube containing a filter followed by two sections of sorbent. The OVS tube was mounted in a plastic protective sleeve which was clipped to the collar of the worker. The filter and sorbent type were selected for each surrogate chemical to maximize sorbent retention and analytical removal efficiency. In the earlier AHETF studies, the analytical laboratories extracted the filter and both sorbent sections together to yield a total residue value (i.e., for particulates and vapors combined). Beginning with the 2008 studies, the laboratories extracted and analyzed the front and back sections of sorbent separately. The front section was analyzed along with the filter and generally contained quantifiable residues. The back section often contained no detectable residues and served as an indication that breakthrough of residues was not a problem during the sampling interval. The front and back residues were then summed to obtain the total amount of chemical collected during the sampling interval. This change was trigged by CDPR expressing concern the lack of breakthrough had not been demonstrated for earlier studies.

For a couple of studies involving solid products and an inorganic surrogate (sulfur), AHETF utilized glass fiber filters in plastic cartridges. Two such filter/cartridge devices were linked in series, so that one served as a back-up sample to check for breakthrough from the first filter. In the laboratory, the front and back filters were analyzed separately and residues were combined for a measure of total surrogate collected.

These air sampling techniques were used for both worker samples and field fortified samples within each study.

For the AHETF-purchased studies, the air sampling techniques varied somewhat and sometimes included different sorbents, filter cartridges (e.g., "IOM Samplers"), or sampling rates (e.g., 1.5 liters per minute).

Suitable Surrogate Chemicals and Products

AHETF and other task forces developed worker exposure data utilizing a number of active ingredients as chemical surrogates to represent the spectrum of registered products. AHETF utilized about 25 surrogate compounds in its exposure monitoring program, including insecticides, herbicides, and fungicides.

The end-use products selected by AHETF were all registered by EPA for the uses monitored and were always used in accordance with the product labels.

The following criteria were used by AHETF in selecting suitable surrogates for use in exposure monitoring studies.

Environmental Stability

Environmental stability was required for surrogates used in exposure monitoring studies. For example, the chemical must not volatilize or degrade during the monitoring period (generally 4 to 10 hours). An unstable chemical might be deposited on the skin during the beginning of a work period, but no longer be present at the end of the workday when AHETF collects the exposure sample. Most agricultural pesticides possess suitable environmental stability that provides pesticidal efficacy (i.e., the chemical must remain on the treated site long enough to exert effective pest control) and stability on worker exposure matrices. All AHETF surrogates were classified as either low (10^{-7} to 10^{-3} mm Hg) or negligible ($< 10^{-7}$ mm Hg) volatility. Stability during test, transit, and storage conditions was confirmed during all AHETF studies by field fortification measurements and storage stability studies. These supporting studies are discussed later in this chapter.

Analytical Properties of Surrogates

The selected surrogate active ingredients met specific analytical requirements that were necessary to allow the data collected using the compounds to fulfill a role of generic data. Analytical methods for all exposure matrices were accurate, robust, and precise with no complications from interferences or background levels. The analytical methods adequately recovered the surrogate active ingredients from worker matrices used for field fortification assessments that included a wide range of chemical loading levels and environmental conditions.

Limits of quantitation (LOQs) for surrogate active ingredients were sufficiently low for worker exposure matrices to ensure quantitation of residues occurring on the various matrices.

Surrogate active ingredients exhibited adequate freezer storage stability. Freezer storage stability studies were conducted on surrogates before they were used in the field. Storage stability was also confirmed during each study by field fortification samples that were stored and analyzed along with the collected samples.

Additional details about the acceptance criteria for existing studies and analytical method development/validation are presented later in this chapter.

Suitability of Inorganic Surrogates

The majority of active ingredients within the scope of the AHETF were synthetic organic compounds, and nearly all AHETF surrogate actives were organic compounds. The single inorganic surrogate active ingredient used, elemental sulfur, was used by AHETF as a representative wettable powder. Considering the fate of pesticide products in the environment, when the chemical is deposited on human skin and adheres to the surface it will either dissipate (i.e., be volatilized, dislodged, chemically altered, etc.) or be potentially available for absorption into the tissues. It was reasonable to assume the use of inorganic and organic surrogate chemicals would result in similar exposures if the surrogate active ingredients had the requisite stability, analytical sensitivity and precision, and were designed to be handled with the identical equipment and in the same manner. Using sulfur for the open pour mixing/loading of wettable powders scenario was discussed and agreed to by the appropriate regulatory agencies (EPA, PMRA, and CDPR) at a JRC meeting.

Additional Considerations for AHETF Surrogates

AHETF selected surrogate active ingredients that were associated with a variety of scenarios or use sites since diversity in handling conditions was a priority. AHETF selected surrogates with relatively high use rates to support the secondary objective (post-rule) to explore the relationship between amount of active ingredient handled (AaiH) and the resulting dermal exposure results. This design element was supported if the range of AaiH was at least an order of magnitude, and high use rate active ingredients make this easier to achieve. AHETF also chose surrogate active ingredients and formulated products that were likely to be used by growers and commercial applicators. This increased the number of employers that might be willing to participate in the study and provided MUs that involved real-world conditions.

Most active ingredients selected had "minimal" PPE requirements. These active ingredients allowed a single layer of clothing to be worn and for most scenarios no requirements to use special engineering controls. This approach resulted in exposure data useful to estimate "protected exposure" for mitigation purposes (e.g., to estimate dermal exposure if a worker wore an additional layer of clothing and/or gloves) by application of a clothing protection factor.

Risk of Exposure to Surrogate Chemicals

Surrogate active ingredients and "end-use" products were associated with acceptable risks to the subjects in studies. The workers recruited for AHETF monitoring studies were professional pesticide handlers and there were no reasonably foreseeable incremental risks to subject workers from handling the surrogate products used in AHETF monitoring studies. The AHETF monitoring program was designed to generate exposure data for workers who handle a pesticide for a period of time representing a single (scenario-specific) work day. Post-rule, AHETF did not use workers for repeated measurements within the same scenario, so acute toxicity was a primary consideration in selection of surrogates for exposure monitoring studies. Because acute effects are formulation-specific, researchers discussed with each study participant the particular effects for the actual test substance they handled. In addition, the surrogate chemical products used by AHETF were registered by EPA for the uses monitored, and were used in accordance with label and WPS requirements. This means that handler exposure levels were not expected to reach a "level of concern" to the EPA since they had previously determined that margins of exposure (MOEs) for the labeled uses and rates, typically based on no-observed-adverse-effect-levels (NOAELs) in animals, were acceptable. Nevertheless, for post-rule studies, AHETF ensured acceptable risks of exposure to surrogate chemicals by re-calculating MOEs for all potential surrogates for all studies. This included MOE calculations for dermal and inhalation exposure, plus an MOE for both routes of exposure combined. Only when all the MOEs were acceptable (again, based on levels of concern specified in current regulatory documents) did AHETF allow use of the particular surrogate active ingredient in its studies.

Design Principles of AHETF's Testing Program

Post-rule, EPA and AHETF re-considered the entire AHETF testing program and many changes were made to how AHETF designed and conducted its exposure monitoring program. The result of those efforts were summarized in a Governing Document that described the need for new human exposure data and provided a detailed set of design principles for how AHETF proposed to conduct its research going forward. The following sections describe the primary considerations that went into preparing the Governing Document as well as the important design principles that were developed to guide the remainder of the AHETF testing program.

Governing Document

In mid-2006, AHETF submitted several study protocols to EPA which were then reviewed by the newly formed Human Studies Review Board (HSRB). HSRB asserted that AHETF exposure monitoring studies could not be determined to be ethically valid until the study protocols are first determined to be scientifically valid. Since the protocols did not include all the measurements needed to complete a scenario, there was no way for HSRB to judge the scientific validity of the individual study protocols and HSRB recommended that EPA reject consideration of research based on these protocols (*8*). In particular, HSRB didn't know how many total workers would be exposed and AHETF did not explain how the resulting data would be analyzed statistically. In addition, HSRB was not satisfied that the ethical concerns with using humans in the AHETF exposure monitoring program were satisfactorily addressed.

This first HSRB interaction resulted in two important EPA actions:

1. EPA scheduled a Scientific Advisory Panel (SAP) meeting to get input on the need for new exposure monitoring data and the suitability of current passive dosimetry techniques for monitoring exposure potential.
2. EPA asked AHETF to prepare a document that summarized its entire testing program, the basic approach that would be used to select subjects, the scientific basis for sample sizes, and a more thorough discussion of procedures that would be followed to protect subjects and inform them of the risks of study participation.

In preparation for the SAP meeting, AHETF summarized the technical aspects of its monitoring program in a document entitled "Technical Summary Document For a Multi-Year Pesticide Handler Worker Exposure Monitoring Program". It primarily discussed the need for additional human exposure data, the statistical sampling design for exposure data, and the statistical basis for how much exposure data may be needed in a scenario. In response to the second item, AHETF produced a document called the AHETF "Governing Document for a Multi-Year Pesticide Handler Worker Exposure Monitoring Program". This "Governing Document" included much of the technical information from the SAP document, but also included extensive discussions about ethical issues such as risk-benefit analyses, recruitment procedures, consent procedures, and methods to protect vulnerable groups. Both of these documents provide extensive information about AHETF procedures.

In 2007, AHETF submitted to EPA and the HSRB a *final draft* of the Governing Document (dated May 22, 2007) and a variety of other documents to support the scientific and ethical validity of the entire AHETF exposure monitoring program. This submission did not include a specific study protocol. The HSRB (*9*) fundamentally supported the ethical aspects of the monitoring program, but recommended adoption of a formal probability sampling approach to obtain existing worker-days instead of the purposive diversity sampling approach proposed by the AHETF. Although it is simplistic and sometimes misleading, probability sampling is often referred to colloquially as 'random' sampling. After that review, EPA further considered the merits of probability sampling versus purposive diversity sampling and determined the latter is more appropriate for the worker exposure data it needed to support pesticide product safety assessments. The primary considerations were that random sampling was not feasible, would require many more MUs, and could not ensure data sets would represent a variety of equipment and procedures, which is the cornerstone of a purposive diversity sampling approach. However, the EPA also concluded that random elements should be incorporated into study designs whenever feasible to reduce the potential for intentional selection bias.

The first complete version of the Governing Document (Version 1, dated April 7, 2008) addressed the comments of EPA and HSRB and detailed the procedures AHETF intended to follow to collect additional pesticide handler exposure data. The Governing Document supported new pesticide handler exposure monitoring study protocols reviewed by EPA and HSRB in 2008 (*10*, *11*). A detailed MU Plan was also developed for each scenario involved in the new studies. The MU Plans included a justification for new data and a rationale for how much data would be generated to complete the scenario (taking into consideration any existing data). Each submitted study protocol referenced the Governing Document and the appropriate scenario plan(s), and provided study-specific information to fully address the scientific and ethical validity of the protocol and proposed research.

External Input for AHETF Testing Program

AHETF sought input on its testing program and procedures from a variety of external sources, the most important being:

- National Academy of Sciences report on intentional dosing studies
- Environmental Protection Agency
- Communities impacted by the research, including the scientific community and the farmworker community

In 2003, the EPA commissioned the National Research Council of the National Academy of Sciences (NAS) to provide recommendations about the acceptability of intentional dosing studies involving pesticides and humans for EPA's regulatory decision making purposes. This report was thoroughly reviewed by AHETF and used as a guide to determine what changes might be appropriate and to help prepare portions of the Governing Document. The primary use of the NAS report was to help AHETF evaluate the risk-benefit analysis which must conclude that the benefits of human studies outweigh the risks to study participants. The Governing Document indicated that AHETF studies could generate societal benefits beyond the minimal benefit of increased knowledge because they could meet the following specific criteria listed in the NAS report: result in more stringent regulatory standards; enable EPA to adopt a public health measure it otherwise could not adopt; support approval of a product that protects public health; and improve the scientific accuracy of risk assessment for a class of chemicals and/or EPA decisions.

The Human Studies Rule impacted EPA's Office of Pesticide Programs in addition to pesticide registrants. In particular, EPA was required to establish the HSRB and needed to establish new policies and procedures for evaluating third-party research studies that were subject to the rule. For well over a year post-rule, AHETF and EPA worked together to evaluate the AHETF testing program and determine what changes were needed in order for EPA to fully evaluate the scientific and ethical validity of exposure monitoring studies. The results of these discussions are summarized in the Governing Document and led to significant changes in AHETF study protocols and Standard Operating Procedures (SOPs). The most important changes were associated with: evaluating risks to participants, risk mitigation procedures, recruitment procedures, equitable selection of subjects, sample size determinations, and statistical analysis procedures. These topics are discussed later in various parts of this chapter.

At the April 2007 HSRB meeting, the board requested that AHETF "broaden participation of the scientific community and parties with direct interest, such as the labor community", who will be "best able to communicate with other workers and assist researchers" (*12*). In response to HSRB's request, AHETF worked with EPA to evaluate what community input might be appropriate. Two communities were identified that would be involved with the individual studies supporting each pesticide handler scenario. First, a resource community that shares AHETF's interests in identifying the scope of a scenario, develops and/or delivers products and services to handlers, and can identify potential farming operations as sources of subjects. This community includes AHETF member representatives, state (e.g., CDPR) and national (e.g., EPA, PMRA) regulatory representatives, state agricultural extension specialists, agricultural equipment manufacturers, commodity / use site organizations (e.g., Cotton Growers Association, Vegetation Management Association) and pesticide applicator associations (i.e., pilot and application company owners and the National Agricultural Aviation Association, NAAA). Second, the handlers themselves as a community with interests in recruitment procedures and the informed consent, with its subsidiary issues of understanding risks and benefits; monitoring procedures; embarrassment; pregnancy testing; privacy/confidentiality of records; vulnerability/voluntariness/coercion; payment; and, adequacy of the consent form.

Post-rule, AHETF involved the resource community during the planning step for new scenarios and such expert input is discussed in detail in a later section named "Expert Input on Program Design". To address the second community, handlers themselves, AHETF and EPA agreed that a sample of various types of handlers should be contacted to review specific task force procedures and documents to obtain their feedback. AHETF conducted seven focus group meetings with handlers across the U.S. from June 2008 to March 2009 designed primarily to obtain feedback from pesticide handlers on the AHETF recruitment and consenting procedures, as well as the content and understandability of the Informed Consent Form used in the studies. Meeting attendees included growers, grower employees, commercial applicators, commercial applicator employees, and owners or operators of agricultural operations who do not directly handle pesticides. In addition, attempts were made to include females and Spanish speakers, however these portions of the community were difficult to enroll.

At the focus group meetings AHETF researchers presented details about specific AHETF procedures (e.g., how AHETF planned to recruit subjects and how exposure is monitored) and standard AHETF documents (e.g., recruitment flyers and consent forms). Input was then obtained from individual participants on specific questions posed in a written questionnaire, including topics such as:

- general recruitment process
- effectiveness of recruitment flyer
- procedures to prevent subject coercion
- could they identify any organization that represents their interests as a pesticide handler
- concern about taking an over-the-counter pregnancy test (female workers only)
- concern about researchers taking photos and videos
- concern about having a medical professional on site
- appropriateness of $100 for subject participation

This input was then used to modify or improve AHETF procedures and documents.

Summary of Changes to Governing Document Version 2

In 2009 and early 2010, AHETF conducted studies that were approved by EPA and HSRB in 2008, most notably six studies designed to complete the scenarios of closed cab airblast application and open cab airblast application. This provided AHETF with in-field experience with the new procedures established by the Governing Document. A second version of the Governing Document was then produced which included some changes to reflect slightly revised practices; however, the basic process for planning scenarios and collecting MUs was fundamentally unchanged. The most important changes reflected in Version 2 of the Governing Document are:

- The risk of exposure to surrogate chemicals and the background risk of agricultural work were no longer considered risks related to the research (and so they were no longer detailed in protocols or consent forms). However, label precautionary statements were still required to be discussed with each subject. This change was the result of comments made by the HSRB at their October 2008 meeting (*11*).
- Changed to reflect AHETF plans to have typically one study per scenario in the future, often changing "study-specific" to "site-specific". This was changed to improve the efficiency of developing and reviewing protocols and reports.
- Changed the heat index cutoff from 120° F to 105° F. This change was suggested by the California Department of Pesticide Regulation as a means of being more protective of workers in hot weather. Although the change in the heat index cutoff requested by CDPR was limited to MUs conducted in the open and closed cab airblast studies, AHETF decided to implement the lower heat index of 105° F for all monitoring areas and all future scenario studies so all subjects would be similarly protected.
- Clarified that one cluster per monitoring area is always the conservative assumption for planning purposes, but that during the execution of a study, the MUs in a monitoring area might result in more than one cluster due to temporal and or spatial separation of the MUs. This might provide greater accuracy but reduced cost-effectiveness.
- Provided a better description of how restricting the similarity of MUs (e.g., one MU per stratum of AaiH, one MU per employer, etc.) ensures diversity. Also established a limit of one MU per employer per monitoring area.
- Added the desire to have more than one recruited worker available per MU as a means of providing some randomness in the final selection of subjects.
- Updated some consent procedures (e.g., no translator and how to find witnesses). Most importantly, this change explained that AHETF will have a bilingual person as a member of the research team who will do the consenting process instead of relying on translators, and also clarified the procedures that will be followed in finding an appropriate witness when necessary (e.g., the subject must identify the witness).

Scientific and Ethical Validity

In late 2006 EPA provided direction on the ethics-related items that must be included with protocol submissions and the "Framework for Protocol Assessment" that listed 64 items that must be explained or referenced in protocol submissions. This information guided the submission of all future AHETF study protocols and reports.

AHETF and EPA then worked together to describe and develop a set of guidelines to ensure exposure monitoring studies could meet these new validity requirements. AHETF discussed all of the criteria in the Governing Document and the following sections discuss important components of scientific and ethical validity for worker exposure studies and how these criteria were met.

Regulatory Need for Additional Data

Pre-rule, AHETF and the JRC jointly determined which scenarios required new exposure data. Post-rule, there was a formal need to justify the ethical validity of exposing human subjects to the risks of study participation before new data could be collected. At the 2007 SAP meeting, the Panel agreed with EPA that "additional data could significantly improve the Agency's ability to assess worker exposure" and agreed with the Agency's concern about the limitations of the existing Pesticide Handler Exposure Database (PHED) (*6*). However, EPA and the HSRB were clear that the need for additional monitoring data must be established for each scenario for which AHETF proposed to generate additional data. Consequently, AHETF produced scenario-specific Monitoring Unit (MU) Plans which discussed the acceptability of existing data that could be identified from the following sources:

- Data already in PHED
- Data available from AHETF member companies
- Data available in public literature
- Data previously submitted to the regulatory agencies

Post-rule MU Plans discussed any existing data that met AHETF acceptance criteria and proposed additional data be generated only when there were insufficient data to adequately describe the scenario in question. This process ensured that workers would not be subjected to study risks unless scientifically justified.

After the SAP meeting, EPA issued Pesticide Registration (PR) Notice 2007-3 (*13*) which announced the formation of AHETF. This PR Notice formally confirmed EPA's opinion that additional worker exposure data were needed and indicated AHETF was formed to generate such data.

Alternatives to Human Monitoring

Another necessary condition for both scientific and ethical acceptability of the AHETF program is that the use of humans to generate new data be justified. This requires that human data are necessary for the regulatory risk assessment process and that adequate alternatives to conducting additional human exposure monitoring are not available. This issue was discussed at the 2007 SAP meeting. Based primarily on the limitations in PHED data, the SAP endorsed the need for new worker exposure data (*5, 6*):

> "The Panel is clearly of the opinion that additional worker exposure data collected on human volunteers under field conditions and label requirements on chemicals that have been approved by the Agency are necessary."

The SAP evaluated the AHETF acceptance criteria for existing data and concluded they were adequate to determine whether those data were appropriate. When insufficient scientifically acceptable data could be identified for a particular scenario, then no viable alternatives to performing additional human monitoring studies exist for generating an updated exposure database. Post-rule scenario plans therefore proposed new human studies when existing data were insufficient.

Defined Scientific Question(s)

Several scientific validity criteria deal with the need for a valid scientific question; a clearly defined purpose; specific objectives and/or hypotheses; and a study design that is suitable to achieve objectives or test the hypotheses. AHETF was formed to answer the scientific question "What is the distribution of worker exposures to pesticides for each occupational pesticide handling scenario in the testing program?" This information is needed by EPA (and other regulatory agencies) to assess risks to workers who handle pesticides.

Sufficient Statistical Design

Post-rule, a very important hurdle for EPA and AHETF was to develop a handler exposure monitoring program that included a suitable statistical design to evaluate the goals and objectives of the AHETF testing program. The statistical design involved two primary components:

1. Defining goals for data adequacy (termed benchmark objectives) and determining a statistics-based sample size (i.e., number of MUs) that would be likely to meet the benchmark objectives; and
2. Specifying in advance how the results of a completed scenario would be statistically evaluated.

Two benchmark objectives were described in the AHETF Governing Document:

1. One primary benchmark objective established a goal of having a certain accuracy, generally 3-fold, for common statistical measures, such as means and upper percentiles, for normalized exposure potential for a completed scenario.
2. A secondary objective was established that evaluated how well a normalizing factor correlates to total exposure for the set of MUs in a scenario.

Sample size determinations were based on a random nested (or cluster) sampling reference model. This model took into account the observation that normalized exposure measurements between MUs conducted close together, both spatially and temporally, tend to be more similar than those much farther apart. Statistical simulations based on the model were then conducted to determine an appropriate configuration of MUs that should meet the two objectives. Simulations took into account the reference sampling model and assumptions for how variable exposure data sets tend to be and how much correlation is expected within a cluster. Scenario MU Plans were then prepared for all post-rule scenarios that required additional MUs and each plan provided a characterization of the scenario, the statistical basis for the number and configuration of MUs, and described the planned diversity selection methodology, and how AHETF planned to analyze the final data. Completed scenario data sets were generally analyzed using the actual variability and correlation measures, plus the actual clustering of MUs, to evaluate how well the primary benchmark objective was met. Actual total dermal and total inhalation exposures for each MU were also compared to their corresponding normalizing factor, generally AaiH, to see how well the secondary benchmark objective was met.

Ethical Design of Program and Studies

Post-rule, AHETF and EPA worked closely to address a variety of specific procedures that had ethical implications and that HSRB would need to review for post-rule worker exposure studies. This process was started when EPA produced a document called "Draft Framework for Developing Best Practices for Recruiting, Screening, Informing, and Consenting Human Subjects of Occupational Exposure Studies with Pesticides" (*14*). This document addressed four overarching concerns that were discussed in terms of what would be necessary to be addressed by study protocols in order to meet the requirements of the Human Studies Rule. Each of these concerns is discussed next in terms of how AHETF addressed the issues when designing post-rule scenarios.

In addition, because of the heightened focus on ethics issues, AHETF adopted the requirement that all researchers who might interact with subjects must obtain ethics training. Acceptable ethics training included successful completion of the course from the National Institutes of Health (Protecting Human Research Participants (PHRP)) and/or the Basic Collaborative IRB Training Initiative Course (CITI; The Protection of Human Research Subjects).

Equitable Selection of Subjects

As a condition for approval of proposed research, the IRB (institutional review board) must determine that subject selection is equitable (or fair). In making this assessment the IRB must take into account the purposes of the research and the setting in which the research will be conducted and should be particularly cognizant of the special problems of research involving vulnerable populations, such as economically or educationally disadvantaged persons. Since there is no direct benefit to subjects participating in a worker exposure study, the concern is for the potential for inequitable reliance on subjects from vulnerable populations rather than excluding potential beneficiaries. Put another way, the essence of equity in selection of AHETF research subjects is that the burden of bearing the risks of research be fairly distributed. EPA identified several important components of fair distribution, including:

- Representativeness of sample
- Appropriate use of inclusion/exclusion criteria
- Special consideration for vulnerable populations
- Appropriate recruiting strategy

Representativeness of sample for AHETF studies included primarily the concern that a small sample of workers that participate in exposure monitoring for a scenario must represent the entire population of workers that might handle pesticides in that way. This led to AHETF scenario design procedures that included defining the target population, obtaining large lists of potential employers, screening employers from random samples from lists, and incorporating random elements in the employer and subject selection process whenever possible. These specific procedures were summarized in the Governing Document and several post-rule SOPs. In addition, scenario-specific considerations were also presented in study protocols and scenario MU Plans. HSRB review of several proposed post-rule scenario studies also led to the requirement to evaluate, post-study, whether the specific employers/subjects actually used in the study were similar to the entire population of handlers in each monitoring area where MUs were collected. When conducted, these evaluations were included in study reports.

AHETF consulted with an external ethicist about many of these issues. Some important decisions were made such as to include only English and Spanish speakers and much discussion took place about specific inclusion and exclusion criteria that were appropriate for AHETF studies. After several discussions with EPA and the HSRB, a complete list of eligibility criteria was presented in the Governing Document. These eligibility criteria were also listed in individual study protocols in the early studies, but were eventually moved to a SOP with the study protocol only addressing study-specific inclusion or exclusion criteria.

The use of vulnerable groups in AHETF studies was discussed in detail in the Governing Document. The following groups that might be included were identified: employees of a grower or commercial applicator; limited or non-readers; poor people; workers without health insurance; and illegal immigrants. How each of these groups would be treated was discussed in the Governing Document and SOPs and several procedures were developed or expanded. These included a standard non-coercion form for employers to sign, a process for assessing the literacy level of potential subjects, procedures for using witnesses for workers with limited literacy, appropriateness of the remuneration offered, justification for reimbursing employers for the cost of pesticides used in a study, procedures for ensuring there will be no cost to subjects, and discussions about why AHETF will not inquire about the immigration status of potential subjects.

And finally, the topic of how AHETF would recruit employers and subjects to ensure equitable selection of participants was discussed thoroughly in the Governing Document and a few SOPs. These recruitment procedures are discussed in detail later in this section.

Fully Informed Choice to Participate

Informed consent is the knowing agreement of an individual obtained without undue inducement or any element of force or coercion. The process involves reading, understanding and signing an informed consent document as well as having a knowledgeable member of the research team discuss the details of study participation. EPA's 'framework' document (*12*) provided an extensive list of necessary components for informed consent forms plus discussions about the consent process, including the importance of determining the competence of the potential subject, handling language differences, minimizing complexity of forms, defining appropriate settings for obtaining consent, communicating risks to potential subjects, and confirming understanding by potential subjects.

While AHETF always had a detailed consent form and established procedures for obtaining consent that had been reviewed by an IRB, there were additional post-rule changes that needed to be made. In particular, many of the standard procedures associated with obtaining consent were documented in the Governing Document and SOPs, standard consent wording was evaluated to ensure an eighth grade reading level, risk descriptions were expanded, the risk-benefit balance was more fully documented, and a new process for assessing whether potential subjects understood all the components of the consent form was developed. In addition, an extensive effort was undertaken to ensure that consent form wording and terminology used was properly translated into Spanish and was expected to be understandable by Spanish-speaking pesticide handlers.

Representative Tasks, Equipment, and Procedures

A major challenge to overcome when estimating exposure for workers who handle pesticides is that several parameters contribute to the likelihood and level of exposure. These include factors such as handling liquids versus solids, product packaging type, open versus closed loading systems, various equipment types, use of personal protective equipment (PPE), and whether the worker mixes/loads or applies or does both. The number of combinations of these parameters makes it impractical to generate data for all situations, so a logical (but discrete) set of handling scenarios needed to be developed that involved important combinations of these parameters and that will be useful for the intended regulatory purpose.

The workers monitored in the studies and handling tasks performed in developing generic data must be representative of the scenario being studied and include the following:

- Activities, equipment and procedures must be well-described and representative of typical agricultural mixing/loading and application practices.
- Individuals monitored must be normally employed in the mixing/loading and/or application of agricultural products.

The following sections describe the fundamental principles used by AHETF when establishing its set of handling scenarios and designing the entire monitoring program.

Discrete Task(s)

The first level of establishing discrete and useful scenarios was to deal with the basic handler tasks of mixing, loading, and application. AHETF and JRC agreed early on that mixing and loading tasks should generally be monitored together, but separately from application tasks since there were too many combinations to do mixer/loader/applicator (MLA) studies for all possible combinations. In addition, mixing and loading are typically conducted by the same worker, but application is sometimes performed by separate workers during the normal course of agricultural pesticide operations. However, it was recognized that some situations are so commonly conducted by the same person that MLA might be more appropriate. These decisions about mixing/loading only vs. application only vs. mixing/loading/application were made on a scenario-by-scenario basis with discussions between AHETF and the JRC. In addition, a few special tasks were identified that warranted separation, such as seed treatment activities, where several discrete tasks resulted in several distinct scenarios.

Formulation Types

Liquids and solids can have widely different exposure potentials and often involve different techniques and equipment. AHETF intended to cover all common formulation types including liquids and solids, and including the use of engineering controls such as closed systems and water-soluble packaging. For most application-only scenarios, the formulation was not always specified but the physical form of the mixture that was applied was specified, for example application of a liquid versus application of a granule (that is not diluted in a liquid). With regards to mixing/loading/application scenarios, AHETF sometimes specified the formulation type and sometimes did not.

Engineering Controls

Engineering controls are designed to reduce exposure potential. For example, the following were considered important engineering controls that warranted multiple scenarios: open pouring vs. closed systems for liquids; open pouring vs. closed systems for granules; open pouring vs. water-soluble packets for wettable powders; and open cab vs. closed cabs for airblast and groundboom applications.

Mixing and/or Application Equipment

In a few cases, specific equipment types were so potentially important to exposure potential, that multiple scenarios were necessary. For example, the rights-of-way applications of liquids where two common application equipment techniques led to two scenarios: handgun versus backpack sprayers.

Baseline Personal Protective Equipment

Personal protective equipment (PPE), such as gloves, aprons, eyewear, respirators, and extra layers of clothing, can have a large impact on handler exposure potential. The basic approach of AHETF was to generate data for the baseline set of clothing and PPE consisting of: single layer of clothing and chemical-resistant gloves. The primary reason for the use of baseline PPE is to provide a "worse case" exposure measurement which is protective of workers. Secondly, regulators are often comfortable applying a protection factor to these higher measured exposure levels that would estimate a lower exposure for a worker wearing extra clothing or PPE. AHETF never requested a subject to wear less clothing or PPE than required by the label(s), or than they would normally wear.

Target Population

Post-rule, each MU Plan defined the target population for the scenario and spelled out how AHETF would "sample" MUs to represent the target population. A handler-day (or HD) is the basic conceptual element of a pesticide handling scenario. Each handler-day corresponds to a particular worker and the scenario-related activities that they perform during a single work day. The scenario (or 'target') population is the conceptual set of all possible HDs resulting from the scenario-specific use of a pesticide. Here, target population is used in its statistical sense to mean those 'units AHETF would like their data to represent'. The distribution of worker exposures over all HDs in the target population would characterize the exposure potential of workers performing the scenario. In practice, the scenario's exposure potential can only be experimentally determined from a set of synthetic handler-days, called monitoring units. Each MU is a worker and a set of scenario-specific handler-day conditions that have been experimentally selected (i.e., chosen, restricted, or scripted) to provide exposure measurements representative of one worker over a typical workday. The set of MUs is considered a sample of handler-days from the target population. Thus, the distribution of worker exposures in this sample can be used to represent the distribution of exposures over the entire scenario population.

Diversity Selection of Research Parameters

In post-rule scenario sampling plans, the proposed MUs were not a purely random sample from the scenario population. AHETF found this to be impractical, if not impossible. Rather, the intent was to obtain MUs spanning a diversity of conditions that were expected to influence exposure, either directly or indirectly. This process was referred to as 'diversity selection' of handler-days. A small set of MUs better described the general magnitude and range of future exposures possible for the entire scenario when the MUs all differed with respect to characteristics expected to have relatively large, but typically unpredictable, impacts on exposure. The most useful of such 'diversifying' characteristics were those known to control multiple factors likely to impact exposure. In general, such meta-characteristics were geographic location, date (e.g., month, season, year), and worker identity (since individuals sometimes perform tasks in their own particular way).

Maximum diversity would clearly result if every MU could be selected to differ from all others in the sample in all important characteristics considered. AHETF considered this impractical for many meta-characteristics. Consequently, AHETF planned a two-stage diversity selection approach to obtain the desired set of new MUs for post-rule scenarios. In the first stage, a limited number of diverse geographic locations, referred to as monitoring areas, were selected. Then, at the second stage, MUs within each monitoring area were selected to be diverse with respect to other characteristics (that were spelled out in the scenario MU Plan).

Because the diversity-selected MUs would not be a random sample from the target population, statistical theory cannot predict the impact of changing the 'sample size' (i.e., the number of MUs and/or monitoring areas). To determine the necessary number and configuration of new MUs needed for a scenario, AHETF adopted a benchmark approach using random sampling as a 'reference model'. Then, the number of new monitoring areas and MUs that would result in desirable statistical properties if the reference model were actually true would be considered a 'benchmark' sample size. The reference random sampling model and benchmark objectives used on simulated data to determine sample sizes for new MU selection designs are described in a section named "Design of Handler Scenario MU Plans".

The following sections describe how MU conditions were diversified in scenario plans.

Workers

Worker identity might impact measured exposure since varying workers could result in different worker practices and/or behaviors. Even pre-rule, AHETF tried to use as many different workers as possible in each study, but for efficiency reasons often allowed some workers to provide more than one measurement. This causes some correlation within a set of measurements that can cause the collection of measurements to be less variable than a set of equal size involving all different workers. Post-rule, AHETF made strict similarity restrictions in study protocols that precluded the use of the same worker for multiple MUs in a scenario.

Employers

Pre-rule, AHETF sometimes used multiple workers from the same employer. However, the concern arose that workers trained by the same company, and using the same techniques and equipment, might result in similar exposure potential. AHETF made strict similarity restrictions in post-rule study protocols that precluded the use of the same employer for multiple MUs within a monitoring area. In a few special cases (such as the right-of-way study where some employers operate in wide areas of the U.S.), the use of the same employer to provide more than one MU was allowed as long as the MUs were in different monitoring areas (although the protocol said this was undesirable).

Location of MU Collection

Location of MU collection might result in correlation within multiple exposure measurements. The concern was that there might be regional differences in how workers perform certain tasks, what equipment are commonly used, or how crops that are treated are grown. Pre-rule, EPA's testing guidelines suggested five MUs from each of three locations, so there had almost always been some attempt to vary the location of MU collection. Post-rule, AHETF continued to specify multiple locations for the MUs within a scenario but tried to choose the locations more deliberately to capture differences in some potentially important condition(s) that might impact exposure potential, such as target crops and how trees are grown for airblast applications.

Timing of MU Collection

Timing of MU collection might lead to some correlation within a group of measurements. For example, seasonal differences in environmental conditions (such as temperature or wind), or stage of growth for crops being treated, might have an impact on exposure potential for some tasks. During the planning stage, this condition was not considered to be very important and scenario plans and study protocols did not require MUs be collected at different times of the year. However, during statistical analysis of post-rule scenarios, MUs collected in the same area, but at different times (for example, 90 days apart or more) were sometimes considered to be in different clusters.

Important Equipment or Handling Procedures

Pre- and post-rule, AHETF spent considerable time determining how handling tasks are actually performed and often designed its study protocols and scenario plans to accommodate various equipment and/or procedures. Post-rule, study protocols became more specific about the desire to vary certain equipment or procedures and were generally presented as (mandatory or desired) similarity restrictions which specified that MUs within a monitoring area should differ between each other in at least one specific way or another. For example, for the closed loading of liquids study protocol it was desired that no two MUs within a monitoring area involve the same system type, transfer set-up, and container size.

Normalization for Generic Use

For prediction purposes, it is useful to express handler-day exposure relative to the value of a normalizing factor (NF). In the AHETF monitoring program, the normalizing factor was always an experimentally measureable quantity that is expected to be proportional to the potential contact the worker has with active ingredient. Potential active ingredient Contact (or PaiC) is defined as the amount of active ingredient that a worker is expected to encounter during a workday. Because each AHETF normalizing factor is only expected to be proportional to PaiC, it should be considered a relative measure of active ingredient contact.

Pre-rule, regulators and registrants assumed that handler exposure was proportional to the amount of active ingredient handled (AaiH). That is, for a given handler situation, the more active ingredient that is handled the more active ingredient will get on the worker's skin or be in the breathing zone. This is the fundamental reason AHETF data are considered to be generic.

Post-rule, AHETF studies continued to formally diversify AaiH and generally became more sophisticated in how AaiH would be diversified so the correlation between AaiH and measured exposure could be examined statistically. For example, it became apparent after analyzing several completed scenarios that when AaiH varied by an order of magnitude there was a good chance to see whether or not the data support the claim that AaiH is likely proportional to exposure. Therefore, several post-rule studies established strata of AaiH that could ensure a 10-fold range in AaiH, both within each monitoring area and for the entire scenario.

Expert Input on Program Design

From the beginning of its testing program, AHETF consulted with a variety of agricultural experts to help plan scenarios. These experts include, but were not limited to, AHETF member representatives, state (e.g., CDPR) and national (e.g., EPA, PMRA) regulatory representatives, state agricultural extension specialists, agricultural equipment manufacturers, commodity / use site organizations (e.g., Cotton Growers Association, Vegetation Management Association) and pesticide applicator associations (i.e., pilot and application company owners, NAAA). The primary purpose of these consultations was to determine how handling activities were actually performed; what equipment and pesticide formulations were commonly used; and to determine whether equipment or practices varied among geographic regions, product types, use sites, etc.

In 2002 (pre-rule), AHETF formed the Agricultural Handler Advisory Panel (AHAP) which included some pesticide industry (i.e., AHETF member company) representatives and external organizations including: Florida Fruit & Vegetable Association, National Cotton Council, National Agricultural Aviation Association, American Nursery and Landscape Association, U.S. Department of Agriculture, American Farm Bureau, Michigan State, Northwest Horticultural Association, and CropLife America. This group had a few meetings or conference calls (minutes were prepared by AHETF) and was instrumental in prioritizing open cab airblast application as a research priority for AHETF. The group also provided technical information about enclosed cabs, backpacks, greenhouse applications, and seed treatment. Some members of the group were also consulted about existing studies that AHETF was evaluating, for example to determine whether particular equipment were still commonly in use.

In 2007 (post-rule), the Technical Advisory Panel (TAP) was developed to solicit information from commodity groups, educators, contract researchers, and university extension personnel to develop specific testing criteria for scenario plans such as for mixing/loading with water-soluble packaging, closed loading of liquids, airblast applications, and closed cab ground boom equipment.

For later studies, AHETF conducted scenario-specific surveys to help design several of its final scenarios. These surveys and follow-up consultations helped AHETF identify application or loading conditions that might impact exposure, and guide the selection of geographic locations and crops /application areas for those scenarios.

Design of Handler Scenario MU Plans

A number of scenarios were either completely or partially populated with 'new' MUs obtained from post-rule, non-observational studies conducted by AHETF. For each of these scenarios, a Monitoring Unit Selection and Construction Plan (MU Plan) was prepared and reviewed by the EPA and HSRB prior to conducting the study or studies. Each MU Plan described the scenario study design and other scenario-specific procedures for selecting MU conditions as described in the Governing Document. They provided a characterization of the scenario, the statistical basis for the number and configuration of MUs, and described the planned diversity selection methodology, and how AHETF planned to analyze the final data. For those scenarios partially populated with MUs previously purchased and/or conducted by AHETF, the MU Plan also described these 'existing' MUs and considered them when determining the number of new MUs needed.

About half of AHETF's 25 handler scenarios involved conducting "new", post-rule MUs. These scenarios involved preparing an MU Plan to guide the generation of additional MUs. The following sections describe the steps taken by AHETF when designing studies to obtain new MUs for a scenario.

Scenario Definition

AHETF developed MU Plans for a variety of handler activities when post-rule data were needed to complete the scenario. These plans began by providing a scenario-specific definition, which was generally based on the formulation, activity, product packaging, and equipment that the data set would represent. In most cases, these basic scenario definitions were the same as, or very similar to, those used in PHED.

In addition to these fundamental definitions, MU Plans also considered scenario-specific attributes that might be necessary for MUs or, on the other hand, that might be excluded from MUs to be collected by AHETF. For example, the airblast scenario plans indicated that the scenario would include any target crop, but all MUs must have involved a foliar application. Thus data were generated for application to vines or trees that had foliage, yet the resulting data are applicable to airblast applications to dormant crops.

AHETF also defined what was involved in the handling task itself. Handling generally refers to mixing, loading, transferring, or applying pesticides. Additionally, it includes the associated tasks of: handling opened containers; disposing of pesticides or pesticide containers; and cleaning, adjusting, handling, or repairing the parts of mixing, loading, or application equipment that may contain pesticide residues. For several scenarios, AHETF specified whether or not these additional tasks could be part of the activities that were monitored. For example, subjects were allowed to rinse non-returnable containers during the closed loading of liquids study. These considerations were also described in study protocols.

AHETF Approach for Obtaining New MUs

The AHETF two-stage diversity selection approach for obtaining new MUs was developed over several years and was based on numerous discussions with agricultural experts, statisticians, the EPA, SAP, and HSRB. This basic approach was described earlier. Scenario-specific modifications to this basic methodology were always discussed in the MU Plans.

Expert Input on Program Design (e.g., committees and surveys)

As discussed earlier, AHETF used a variety of techniques to obtain input from outside experts when preparing MU Plans. Pre-rule, the input from experts was only informally documented. For example, input was documented in minutes of AHAP meetings or with internal AHETF summaries used for discussions at AHETF meetings. For all post-rule scenario plans, however, a detailed summary was presented in the plan of what efforts were made to obtain external input and what information was obtained from these consultations. MU Plan information ranged from call notes for discussions with individual experts to full reports of formal surveys conducted by AHETF.

The MU as a Synthetic Handler-day

Both existing and new MUs consisted of a set of mixing/loading, application, or mixing/loading/application conditions (including the particular worker) that were intended to represent the scenario activities for a single workday. In many cases these MUs were selected from "naturally occurring" handler-days. This meant that monitoring took place for actual workers that planned to handle a chosen pesticide product in a certain way. However, for new MUs, the selected handling conditions were sometimes modified or scripted slightly to ensure that the complete sample of MUs reflected the diversity in the entire population of future handler-days. As a result, some MUs were selected for conditions that were less typical for the active ingredient being monitored, provided these conditions were expected for other active ingredients. Therefore, few MUs were technically "sampled" from a population of scenario handler-days. More correctly, they were viewed as synthetic handler-days derived from both selected and constructed conditions.

Normalized Exposure

For scenario design purposes, it was necessary for AHETF to pre-specify a normalizing factor most likely to be used for the scenario. Then, the anticipated NF (and exposure normalized by NF) was assumed in the MU Plan to determine sample size, design diversity selection methods, and specify the planned data analyses. In the design of all AHETF studies to obtain new MUs, the amount of active ingredient handled (or AaiH) by the MU was the anticipated normalizing factor and most MU Plans specified a stratified diversity selection approach for AaiH.

Diversity Selection of New Monitoring Units

As discussed earlier and in the AHETF Governing Document, post-rule scenario plans were designed to obtain a set of experimental handler-days (i.e., MUs) spanning a diversity of conditions that are expected to influence exposure, either directly or indirectly. Almost all post-rule scenario plans utilized a two-stage diversity selection approach. First, a limited number of diverse geographic locations, referred to as monitoring areas were selected. Next, MUs within each monitoring area were selected to be diverse with respect to other characteristics (as described later).

MU Plans specified the monitoring areas to be used, typically consisting of one or more convenient political entities, such as states (or portions of states) or EPA growing regions. MU Plans also provided the rationale for monitoring area selection, which usually took into account where the handling activity of interest was likely to occur and/or where particular crops were likely to be grown or other use sites exist (e.g., right-of-way sites). For several scenarios, AHETF used National Agricultural Statistics Services (NASS) data to identify areas where certain crop types are grown or identify areas where certain surrogate chemicals or pesticide products are used. Monitoring areas more likely to be selected were those that had a sufficient number of employers and, therefore, more potential workers to recruit. NASS data also helped with this assessment.

Statistical Basis for Number of Monitoring Areas and Monitoring Units

Technically, desired sample sizes (i.e., the needed number of new monitoring areas and MUs per area) could only be determined if the MU selection process was random. Otherwise, there is no statistical theory that predicts how increasing the number of MUs reduces error. The diversity selection procedure described earlier is primarily non-random, although, it might include some random choice components (e.g. contacting potential employers in random order). In such common data collection situations, sample size can only be determined by using a hypothetical random situation as a reference model. A random reference model was constructed so that it reflected important aspects of the two-stage diversity selection process. A process was also developed that addressed the statistical basis for determining an appropriate sample size based on the model that was likely to meet two benchmark statistics. These processes are briefly described next.

Reference Random Sampling Model

For AHETF scenarios, random nested (or cluster) sampling was used as the reference model for the combination of purposive and random two-stage diversity selection actually used for new MUs. In particular, the reference model used in simulations to determine sample sizes assumed that:

- Normalized exposure is lognormally distributed with a known geometric standard deviation (GSD). This also implies that non-normalized exposure is both lognormally distributed and directly proportional to NF, the anticipated normalizing factor for the scenario. In all MU Plans, AHETF used NF=AaiH.
- Normalized exposures between MUs conducted close together, both spatially and temporally, might be more similar than those much farther apart. In simulations to determine sample size, such MUs were modeled as correlated clusters of MUs. This was referred to as intra-cluster correlation (ICC).

For scenarios with no existing MUs, the reference model assumed that there would always be one or more monitoring areas and the same number of MUs per area. Simulations then examined different combinations of number of areas and MUs and assumed that all MUs in the same area belonged to the same cluster with the same ICC. For some scenarios this reference model had to be modified to include some existing MUs, as well as new MUs. The reference model again assumed that all new MUs in the same monitoring area were in the same cluster with the same ICC; however, whenever the existing MUs had a more complex correlation and clustering structure, the reference model was modified to reflect this.

To fully specify the reference random sampling model, values for the variation (GSD) and correlation (ICC) parameters were needed. Based on analyses of exposure from a number of available monitoring studies, Appendix C of the AHETF Governing Document derived a default structure consisting of a GSD of 4 and an ICC of 0.3. Generally, AHETF and the JRC agreed that because there was no additional evidence to suggest otherwise and no other strong opinion to the contrary, these default values of GSD and ICC would be used for determining sample size in many of the AHETF scenarios. However, the Powered Handgun Spraying in Horticultural Facilities (GH/N) scenario was expected to be quite heterogeneous and a GSD of 5 was considered more appropriate in the reference model.

Benchmark Objectives

All post-rule scenario designs used the following two default benchmark objectives (for dermal exposure only) in simulations used to determine sample size.

Primary Benchmark for Normalized Dermal Exposure:

- When the reference model is used to simulate normalized dermal exposure data
- then the correct sample size will give simulated results having calculated geometric mean (GM), arithmetic mean (AM), and 95th percentile (P95) values accurate to within 3-fold of their true (i.e., reference model) values
- at least 95% of the time

Secondary Benchmark for Dermal Exposure:

- When the reference model is used to simulate non-normalized dermal exposure data (and NF values)
- then simulated data with the correct sample size will result in a statistical rejection of the hypothesis that exposure is independent of NF (but will not reject proportionality)
- at least 80% of the time (i.e. the statistical test has at least 80% power)

The primary (or 'most important') objective relates to the desired relative accuracy for key statistics commonly used to characterize the magnitude of normalized dermal exposures. It is considered to be the more relevant objective in order to obtain a set of diverse MUs used to characterize the general magnitude and extent of normalized exposure in an actual handler scenario. EPA provided guidance to AHETF on the minimum degree of accuracy desired for regulatory use in particular scenarios. The consensus was that estimates of the geometric mean, the arithmetic mean, and the 95th percentile of dermal exposure generally were accurate to within approximately 3-fold of the actual population value. Therefore, AHETF and the JRC agreed 3-fold accuracy was an acceptable benchmark for most scenarios. The agreement between AHETF and the JRC of 3-fold accuracy was based on an assessment of the cost and time required to collect MUs following various combinations of monitoring areas and MUs/area.

The secondary (or 'less important') benchmark objective just states that the number of monitoring areas and MUs also be sufficient for the simulated data to provide 80% power for a limited examination of the relationship between exposure and the anticipated NF. This secondary objective was usually satisfied when the primary objective was met as long as the NF had adequate variation within each monitoring area. As discussed later in this section, all MU Plans included diversification of AaiH within each monitoring area as a formal part of the diversity selection process. However, this objective was considered less important than the primary objective and was accommodated only when it did not negatively impact the diversity selection process or the primary objective.

Determining Sample Size

The simulation methodology used to calculate sample sizes when the reference model used was two-stage cluster sampling from a lognormal distribution. Several statistically equivalent configurations of number of monitoring areas and number of MUs per monitoring area were found that met the primary 3-fold accuracy benchmark objective. These equivalent configurations are listed in Table 1.

Table 1. Equivalent Configurations of Monitoring Areas and MUs per Area

Number of Monitoring Areas	*Number of MUs per Monitoring Area*
15	1
9	2
7	3
5	5
4	8

For scenarios with some existing MUs, or for those not using the default reference model parameters (i.e., GSD=4, ICC=0.3), simulations were used to determine accuracy or power for each number and configuration of MUs considered. The sample size(s) that had simulated data satisfying both benchmark objectives were then used in the scenario design.

The simulations that resulted in the five configurations described earlier considered only the primary benchmark objective. However, when a particular sample size met the primary 3-fold accuracy benchmark objective, the secondary benchmark objective was also met provided: (1) the practical range of the normalizing factor was at least an order of magnitude and (2) there is adequate within-monitoring-area variation in the normalizing factor. Therefore, scenario plans included a special similarity restriction on the normalizing factor to increase its variation within each monitoring area.

Choosing a Configuration of MUs within Monitoring Areas

Each MU Plan specifically chose a configuration of areas and MUs that was expected to be as cost and time effective as possible. This was desirable since costs escalate rapidly when a research team makes several visits to a location in order to monitor the desired number of MUs (e.g., five MUs per monitoring area). Ideally, the chosen configuration would meet the following criteria:

- A set of growers/contractors can be identified that are relatively close together in the monitoring area,
- Separate workers expected to make applications within a narrow time frame can be identified from this set of growers/contractors,
- The resulting set of MUs will provide diversity in factors expected to impact exposure.

Eleven post-rule scenarios involved formal MU Plans and they utilized a variety of MU configurations ranging from 3 to 10 monitoring areas per scenario with 3 to 5 MUs per monitoring area.

Diversity Selection Characteristics Considered in Scenario Design

In the second stage of diversity selection, the set of MUs collected *within* each monitoring area should also be diverse with respect to key characteristics suspected of influencing worker exposure. Such diversity would act to minimize within-cluster correlations and improve the amount of information per MU. For scenario designs, only a few characteristics, primarily meta-factors, were formally considered for diversity selection. Meta-factors are characteristics that directly or indirectly influence a number of other factors. Geographic region is a meta-factor addressed by the first stage of diversity selection and worker is an example of a meta-factor addressed in the second stage. Substituting one worker for another alters a number of factors (e.g., behavior, physical stature, stamina, and equipment available) that might affect exposure.

In every scenario design AHETF identified those potential MU conditions that should be formally or informally diversified within monitoring areas. This evaluation was made primarily by exposure assessment experts from AHETF member companies in consultation with exposure assessors from North American regulatory agencies. MU Plans then identified procedures to diversify these conditions. Formal diversification involved similarity restrictions which were well-defined selection rules designed to limit the degree of similarity between MUs. Similarity restrictions were either mandatory or preferred. Preferred similarity restrictions were still well-defined and desirable, but were not required for every MU. Informal diversity selection simply meant identifying potential characteristics that might be varied in some informal manner when selecting MUs from the same monitoring area. In all cases, these diversity restrictions from MU Plans were reflected in subsequent study protocols that guided the collection of specific MUs. Some of the most common similarity restrictions from MU Plans are discussed next.

Similarity Restrictions on Repeat Use of Workers and Employers

MUs obtained from existing studies reviewed by AHETF commonly exhibited the repeated use of the same workers, that is, there were more MUs than individual subjects. This collection of multiple MUs from a single participant, and often from the same employer, was common among pre-rule studies. Because dermal exposure is highly correlated with individual workers, these existing data do not provide as much information as would the same number of new, independent MUs. In addition, AHETF acknowledged the potential that employees trained by and working for the same employer may perform the handling activity in a similar manner; therefore, AHETF, in agreement with EPA and HSRB, specified that no two MUs obtained for the same scenario can utilize the same worker and that no two workers in the same monitoring area used for the same scenario can have the same employer.

Similarity Restriction on the Amount of Active Ingredient Handled (AaiH)

The amount (i.e., pounds) of active ingredient handled (AaiH) was the anticipated normalizing factor considered by AHETF in all scenario designs requiring new MUs. Thus, for design purposes, AHETF also assumed that future users of the scenario data might wish to verify that a proportional relationship between dermal exposure and AaiH is reasonable. This essentially means that efforts should be taken to generate data in as wide a range of AaiH as practical within each monitoring area.

In addition to its potential direct relationship to exposure, the amount of active ingredient handled is also viewed as a meta-factor affecting parameters such as spray volume, number of loads prepared and/or sprayed, etc. Thus, diversification of AaiH induces diversification of such associated factors as well. Therefore, in addition to having a wide range, the MUs within the same monitoring area should not have similar AaiH levels.

A practical AaiH range was first defined for the entire scenario which included AaiH levels that can be readily found in practice and are expected to produce detectable exposure results. AHETF has developed a practical range of AaiH for each scenario, taking into account such factors as the typical use rates of products, types of products available on the market, area that can be treated in a day, etc. This practical range was then partitioned into several strata of approximately equal width on a logarithmic scale, most commonly the same number of strata as MUs per monitoring area. Then, for each monitoring area, there was an attempt to restrict the MUs to no more than one per AaiH stratum. This similarity restriction imposed AaiH diversity on the MUs, ensured a wide range of AaiH levels within each monitoring area, and was designed to result in AaiH levels for all MUs being approximately uniformly located over the entire log AaiH range when log exposure was regressed against log AaiH.

Miscellaneous Similarity Restrictions

In addition to the similarity restrictions on monitoring areas, workers, and AaiH, diversity among MUs was also enhanced by imposing other mandatory or preferred similarity restrictions on each set of MUs in the same monitoring area, when appropriate. These miscellaneous similarity restrictions include the following examples; equipment used, application site, formulation, product container type and size and method of mixing product.

Active Ingredient Selection

Scenario plans identified suitable surrogate active ingredients based on the scenario-specific needs for certain formulation types, packaging types, crop/site uses, etc. All surrogates used were also consistent with the acceptance criteria for surrogates detailed earlier in this chapter. MU Plans also presented the margin of exposure calculations that demonstrated acceptable risks for all surrogate chemicals selected for the scenario, plus citations for all the regulatory documents that support those calculations.

Planned Data Analyses

Each MU Plan described the statistical analysis that AHETF would conduct on the final exposure data set and then summarize in a scenario monograph. The purpose of the statistical analyses was to assess the quality of the scenario's exposure data with respect to the objectives used to determine the desired number and configuration of new MUs. The MU Plans always stated that the AHETF analyses would be based on the following calculations:

1. For each MU, calculate total dermal exposure and total inhalation exposure normalized by the factor assumed when determining sample size. As noted earlier, this factor was the amount of active ingredient handled (AaiH) for all scenarios needing AHETF-conducted studies to obtain new MUs.
2. Calculate those normalized exposure statistics (and confidence intervals) considered for simulated data when evaluating the primary benchmark objective. These would be the data-based estimates of lognormal variation (GSD), ICC, and magnitude (GM, AM, and P95). For each of the three normalized exposure magnitude statistics, calculate fRA_{95}, the upper 95% confidence bound on their true, but unknown, fold-relative accuracy.
3. Using the same log-log mixed-model regression used for simulated data when considering the secondary benchmark objective, determine if the data are consistent with a proportional relationship between exposure and the normalizing factor used. Use the width of the regression slope confidence interval to assess how well the data would have been able to statistically distinguish a proportional relationship from independence, if one or the other were true.

MU Plans indicated the statistical analyses of exposure would account for the expected lognormal distribution and potential correlation due to MU clustering.

Consistency with Benchmark Objectives Used to Determine Sample Size

Each MU Plan also described how to use the statistical analysis results to assess the extent to which actual exposure results are consistent with the benchmark objectives used to determine sample size needed for the scenario. Although confidence intervals for normalized exposure statistics and statistical tests of the exposure versus normalizing factor relationship are usually the most relevant metrics for the completed data set, the planned statistical analyses described earlier can be used in a simple way to judge if (or how well) the data appear to be consistent with the two benchmark objectives. The value for upper fRA_{95} can be compared to the target value of 3.

Assessing consistency with the secondary benchmark objective was similarly straightforward. As discussed in the Governing Document, if the log-log regression test has exactly 80% power to distinguish a proportional from an independent relationship between exposure and the normalizing factor, then the width of the 95% confidence interval for the true regression slope would be 1.4 on average. Therefore, a set of exposure data giving a regression slope confidence interval of 1.4 wide or less is considered consistent with the secondary benchmark objective.

The sample sizes for each scenario were determined using a reference random sampling model and benchmark objectives for *dermal* exposure. However, the MU Plans always stated that AHETF would conduct the same statistical analyses and benchmark consistency evaluation for inhalation exposure. These inhalation exposure results were also summarized in each scenario monograph.

Design of Studies

All studies for which AHETF generated dermal and inhalation exposure data followed protocols developed in compliance with the EPA Federal Insecticide, Fungicide, and Rodenticide Act (FIFRA) Good Laboratory Practice (GLP) Standards (40 CFR 160). This included adherence to applicable AHETF and/or field facility SOPs. They were also based on EPA's guidance documents for dermal and inhalation exposure measurement under Series 875: Occupational and Residential Exposure Test Guidelines (*15*).

AHETF developed and used a standard template for protocols for studies before the Human Studies Rule, which were reviewed and approved by an independent IRB and appropriate regulatory agencies (EPA, PMRA, CDPR). Protocols for studies after the Human Studies Rule (post-rule), however, also required review by the Human Studies Review Board (HSRB) before final IRB approval. As a result, protocols were expanded to reflect those reviews and recommendations made by HSRB in accordance with 40 CFR Part 26. To accommodate these changes, an updated protocol template was developed and followed.

The following sections discuss important components of study designs and protocols, including specification of what scenario(s) is being covered, ethical considerations in place to protect workers, monitoring area definitions, recruitment procedures, and procedures for obtaining acceptable configurations of monitoring units (MUs).

Scenarios Covered

Study protocols always identified the scenario(s) for which MUs could be collected within the study. Pre-rule protocols addressed one or more scenarios, but were not always specific about exactly how many MUs would be collected. Some early protocols were written to allow both mixer/loaders and applicators to be monitored in the same study. For example, a pre-rule study protocol was written with the primary purpose of monitoring mixing/loading of dry flowable formulations. However, the secondary purpose of this study was to monitor workers, if available, who would then apply those spray solutions using scenario-specific application equipment or techniques such as an open or closed cab tractor making a ground boom application.

Post-rule, AHETF began developing a protocol to cover each monitoring area within a scenario. For example, if the scenario involved five monitoring areas, then there were five corresponding protocols. It was later determined that this approach was inefficient and generated unnecessary paper work since the protocol requirements were essentially the same for all monitoring areas within a scenario. As a result, it was agreed that one protocol would cover all monitoring areas within a scenario. In addition, there were two situations in which a single protocol covered the same monitoring areas but pertained to two distinct scenarios. One was the rights-of-way (ROW) protocol which covered the backpack (ROW-BP) and handgun (ROW-HG) scenarios since research showed that applications using these two techniques were likely to take place in similar locations using similar cooperators and similar surrogates. That meant there were built-in efficiencies for generating data for these two scenarios, such as developing only one set of call lists, calling potential cooperators only once, or coordinating the timing of MUs so that the study team could cover both scenarios on the same visit. The other case is closed system liquid loading (CSLL), in which the protocol covered using either returnable or non-returnable containers (CSLL-R and CSLL-NR).

Ethics Considerations for Worker Exposure Monitoring

The primary authority to oversee ethics issues for AHETF studies (both pre- and post-rule) was an independent ethics review board. Post-rule, however, science- and ethics-related recommendations from the HSRB were also considered before final protocol approvals were given by the IRB. Another post-rule ethics consideration was that any researcher working on behalf of the AHETF and interacting with study participants was required to complete one or more training courses for the protection of human subjects. Recertification was obtained every three years, and copies of the certificates of completion for the ethics courses were submitted to the IRB periodically (as general informational submissions, not as study-related submissions).

The following sections focus on AHETF documents and issues as they relate to ethics associated with human studies, including discussions about AHETF's interaction with IRBs, EPA and HSRB.

Institutional Review Board (IRB) Review

Certain documents used in an AHETF-conducted study, including protocols, Informed Consent Forms (ICFs), and any materials distributed to workers such as recruiting materials, were reviewed and approved by an IRB for ethical compliance and submitted to EPA as required by 40 CFR 26.1125. The report of the completed research was subject to 40 CFR 26.1303 requirements to document the ethical conduct of the study.

In all AHETF studies, the IRB had to determine that certain requirements were satisfied before approval was given for the research to proceed (40 CFR 26.1111), including:

- Risks were minimized;
- Risks were reasonable relative to anticipated benefits;
- Subject selection was equitable;
- Informed consent was sought from the prospective subject;
- Informed consent was appropriately documented;
- Data monitoring was done to ensure safety of the subject (medical management plan); and
- Privacy of subject was protected and data confidentiality was maintained.

Study Protocol and Standard Operating Procedures (SOPs)

All study protocols were prepared by AHETF and submitted to an IRB for approval. Pre-rule protocols were reviewed by EPA before IRB approval. Post-rule protocols were first approved by IRB, and then submitted to EPA and HSRB for review. Any changes were incorporated and sent back to IRB for a final approval. The exception was that observational study protocols did not require HSRB review. For studies conducted in California, protocols were first reviewed and approved by the California Department of Pesticide Regulation (CDPR). Any changes requested by CDPR were incorporated into these documents before they were submitted for review and approval by the IRB.

Pre-rule, one of the first activities of the AHETF was to develop a set of Standard Operating Procedures. Unlike protocols, they did not require IRB review and approval. In June 2006, HSRB met to review AHETF protocols for the first time (*8*). Based on HSRB comments/recommendations from the June 2006 meeting, AHETF and EPA worked together over the next year to create a new set of SOPs. The next HSRB review of AHETF documents was in June 2007 and the HSRB indicated that the revised versions appeared to be reasonably complete, but made some additional recommendations for revising the SOPs (*16*). Some important SOPs developed post-rule involved ethics-related issues, including the following:

- Identifying risks and mitigation procedures (e.g., heat-related illness)
- Developing procedures for identifying eligible employers through which employees could be recruited
- Recruiting study volunteers
- Developing procedures to accommodate language considerations and requirements

SOPs continued to be periodically updated or created over the course of the field monitoring program, resulting in corresponding changes to the protocol template.

Informed Consent Form (ICF)

Pre-rule and post-rule ICFs were reviewed and approved by an IRB and were prepared in both English and Spanish. Initially, AHETF created and used a generic template. The template was modified as appropriate to include information specific to the study design and/or test substance. Pre-rule and post-rule ICFs contained the same required elements (40 CFR Part 26.1116 - General Requirements of ICFs), but the post-rule ICF was much more comprehensive. Various template versions were prepared over the course of post-rule study conduct to accommodate the many recommendations made by HSRB, EPA and the IRBs. ICFs and the consent process are more fully described later in this chapter.

Recruitment Documents

Pre-rule, AHETF did not prepare specific recruitment documents. Recruitment was primarily done through the use of a Contract Research Organization (CRO) in each monitoring area. These were agricultural researchers familiar with crops and agronomic practices in their regions of the country. Given this, along with extensive contacts with farms in the region, these researchers were able to find potential participants that met the desired study design criteria. Refer to a later section named "Recruitment Procedures" for a detailed description of this process.

Several post-rule documents were developed and used in the recruiting process and each of them required IRB approval. Two of these were prepared in both English and Spanish. One was the Recruitment Flyer which was given to interested employers and employees. It described important features of the study, including the need for study volunteers. The other was the Employer Cooperation Statement, a form signed by the study volunteer's employer indicating that the employer would not encourage or discourage the employee's participation. The signed form was provided to employees during recruitment.

AHETF prepared other post-rule documents which were used as recruitment tools, although not provided in Spanish. Two were sent as a packet to all qualified growers who were identified during the initial screening process. One was a study-specific brochure which described the AHETF, its members and a brief overview of the study being conducted. The other was an introductory Letter to Cooperators which stated that they would be contacted by an AHETF researcher to provide additional information about the study. A small gratuity of $5 was included in the packet for the grower's time to review these documents. This procedure proved to increase the number of positive responses when identifying potentially eligible employers.

Product Labels/MSDSs or Product Risk Statements

The IRB did not review product labels or Material Safety Data Sheets (MSDSs) for content but were provided copies if requested.

Pre-rule, MSDSs and labels were made available to the study team and the label safety requirements were explained to volunteers.

Post-rule, AHETF developed a document called the Product Risk Statement (PRS) which required approval by the IRB. PRSs summarized important information about a surrogate product(s) that was expected to be used in a study, and was attached to the consent form. Information included: product name, packaging, approximately how much would be handled, required PPE, safety recommendations, and potential health effects. EPA later decided that PRSs were not necessary since only commercial products registered with EPA were used in the studies. Instead, the ICFs were revised to provide similar information such as the names of active ingredients that might be used in the study and clothing/PPE requirements.

Medical Management Plan

Specific risks were not addressed in pre-rule protocols; the ICF indicated only that heat-induced illness was possible due to an extra layer of clothing. Although other health risks were identified, heat-related illness was considered to be the risk of most concern because the worker wore an extra layer of clothing.

Post-rule, an extensive heat stress management plan was developed and put into place to mitigate this risk. This plan was communicated with the IRB.

Spanish Translations

To be eligible to participate in AHETF studies, prospective subjects were required to understand either English or Spanish (except the aerial application study since all pilots must speak English). As a result, relevant documents were made available in both languages. These documents were the Recruitment Flyer, the Informed Consent Form, the Employer Cooperation Statement, the Request to Receive Personal Results and the PRS (until discontinued).

Pre-rule, the IRB contracted a certified translator to translate these documents. Post-rule, a bilingual researcher familiar with AHETF studies and agricultural terminology prepared the Spanish version of the ICF based on the English version. The Spanish version was then reviewed and certified by an IRB-contracted certified translator. In addition, AHETF contacted bilingual agricultural experts, such as pesticide safety trainers, and asked them to review the ICF and recruitment flyer templates for simplicity, accuracy and appropriateness of terminology. This project resulted in simplifications to the ICF and confirmation there were no concerns with people speaking different dialects understanding the forms.

Inclusion Criteria

To be eligible to participate in an AHETF study, all potential subjects had to meet certain standard eligibility requirements (i.e., inclusion criteria). Inclusion criteria for pre-rule studies included many of the same components as post-rule (e.g., experienced in the activity, be in good health, not be pregnant, and be at least 18 years old), but post-rule criteria were more comprehensive. For example, the requirement to understand either English or Spanish was added, and several criteria were expanded, such as providing proof that the volunteer was at least 18 years old.

Remuneration of Subjects

This topic was not addressed in pre-rule study protocols, but the consent form informed the subject that they would be paid $100 for each day of participation. Post-rule, subjects were given $20 for attending the consent meeting (whether they participated in the study or not) and $80 for actual participation. EPA agreed this was sufficient to compensate subjects for their inconvenience, but not so large as to create an undue incentive to participate. Subject remuneration was explained in study protocols.

Risks to Subjects

Pre-rule protocols did not provide a detailed discussion of study-related risks. They only stated that subjects would be provided information about potential risks, and these were briefly covered in the associated ICF. Mitigation was not addressed as such; however, there were procedures provided in the protocol that served to reduce some of the risks associated with study participation.

Post-rule study protocols (and consent forms) and SOPs specifically listed study-related risks and provided details on how they would be mitigated. Study-related risk included; heat-related illness, scripting field activities, psychological risks, exposure to surfactants (during hand wash and face/neck wipe), and exposure to chemicals (test substances).

Benefits

Benefits to study participants or the employer were not specifically addressed in pre-rule protocols. However, the employer did receive product without charge for agreeing to participate in the study. For some earlier studies, AHETF decided what surrogate product would be used in a particular study. The product was ordered in advance and the employer then agreed to use it. Later, however, the employers chose what surrogate product they wanted to use and were reimbursed by AHETF.

Post-rule, AHETF identified study benefits relevant to worker exposure studies and incorporated them into study protocols. These included:

- There were no personal benefits to the study participants.
- Cooperating employers were reimbursed for the pesticides used for the study. This was considered a minor benefit when compared to the costs of running their businesses.
- Data from the AHETF exposure monitoring program had the potential to improve the ability of EPA and other regulatory agencies to more accurately assess occupational risks associated with important agricultural handling scenarios. The knowledge obtained from AHETF studies was generalizable and contributed to assessments of the risks of both new and existing pesticides.
- Since there were insufficient existing data suitable for use in a generic database describing the exposure of workers who performed various agricultural handling activities, society likely benefited from data generated by AHETF through improved risk assessments by EPA and other regulatory agencies.

Risk/Benefit Balance

The risk/benefit balance was not discussed in pre-rule protocols but was addressed in post-rule protocols. The following two paragraphs became standard language in the protocol:

By monitoring exposure to professional handlers who follow their normal practices but wear an additional layer of clothing (as an inner dosimeter which traps chemical that penetrates the work clothing), AHETF studies presented a greater than minimal risk to participants. Study participation increased the risk of heat-related illness.

The likely benefit to pesticide handlers as a whole and to society in general, in the form of more accurate measurements of potential exposure to pesticides, was weighed against the risks to participants. As a result of AHETF research, growers and other users have additional options for meeting their individual needs. Therefore, exposure data generated by AHETF and meeting contemporary standards of reliability and quality have likely provided a significant benefit to society.

Because margins of exposure for AHETF studies were acceptable for the products proposed for use, subjects were very unlikely to experience acute toxic effects, and because extensive procedures were in place to minimize these and other risks to participants, the likelihood of serious adverse effects was very small. In summary, AHETF believes the risks to study participants from participating in its research program were reasonable in light of the likely benefit to society of the knowledge to be gained.

Respect for Subjects

For AHETF studies, respect for subjects was addressed by assuring their privacy and their ability to withdraw from a study without consequence. Pre-rule study protocols addressed privacy only by indicating that dressing/undressing would occur in a private area. Post-rule protocols provided SOP references about the procedures employed to help assure subject privacy during recruitment, consent, study conduct, and maintenance of study records. These procedures included; access to personal records or results, recruitment meeting without employer, private room for pregnancy testing & no requirement to provide results, and private consent discussion. Protocols also stated that the consent form summarized important confidentiality issues for subjects, for example protecting confidential records containing personally identifiable information.

The right for a subject to withdraw from research without consequence was always considered one of the most important aspects of protection of human subjects. Pre-rule protocols did not address this topic except to state that the ICF informed the subject of their right to withdraw at any time and for any reason without jeopardizing the job. Post-rule protocols explained that prospective and enrolled subjects would be informed of their right to withdraw without consequence prior to and during the conduct of the research. Furthermore, if a subject withdrew while being monitored, the long underwear and air sampling pump would be removed, and the hand wash, and face/neck wipe samples would be collected if the worker agreed.

Informed Consent Form and Consent Process

The Informed Consent Form is a document that explains in detail the aspects of a study that are important to a prospective subject. The following sections present a summary of the primary components of AHETF consent forms. Italics indicates information that was added or changed as a result of the Human Studies Rule. Following the list of consent form components are discussions of ethical considerations related to the <u>process</u> of obtaining consent for AHETF studies.

Introduction and Purpose: Introduced AHETF; explained study purpose, need for *professionals* to participate, and number of participants; emphasized participation was voluntary, *but participation required they read and sign the consent form; promised subject a copy of everything signed*

Research Participant Selection: Specific list of inclusion criteria, including: non-pregnant, minimum age, minimum experience, good health, and *not employee of a pesticide company*

Study Duration: Stated range of expected duration of monitoring in hours, *plus "on just one day"*

Study Procedures: Separate sections for "before" and "on day of" the study discussed the same procedures as pre-rule, in slightly more detail, and included: *procedures to maintain privacy and confidentiality during pregnancy testing and from photos and videos*

Product Handled: *Showed a list of possible active ingredients and indicated product will be registered for use, chosen by employer, and identified before participation; also that a researcher will review the label* *with the subject, including how much product might be handled, the precautionary health statements on the label, required clothing/PPE, and other safety precautions*

Risks and Discomforts: Listed specific risks: heat stress *(including list of early signs and symptoms), eye or skin irritation from surfactant,* getting sick from exposure to the pesticide product, discomfort from wearing air sampling pump, embarrassment during dressing and undressing, *concern about taking a pregnancy test, accidental disclosure of personal information from photos/videos, and working longer than normal because of the extra time it takes to collect samples for analysis*

Injury to Participants: Medical treatment will be available *at the workplace* or at an identified nearby health care facility, AHETF will cover the cost of study-related medical expenses *not covered by private insurance, and medical treatment records will not be part of the research records; also participant may refuse treatment unless the medical professional decides the participant is too sick to make a decision about getting medical treatment*

Confidentiality: *Subject name will appear only on signed consent form and the optional form requesting personal study results, records with the subject's name will be stored in a secure location with limited access,* subject will be identified by a code everywhere else, no information will be provided to the subject's employer, and complete confidentiality cannot be guaranteed because a report will be sent to governments

Costs: There will be no costs for participation

Benefits: No direct benefit for study participation, but information will help protect workers; *also employer might benefit if reimbursed for the product used*

Payment for Participation: $20 for consent meeting and $80 for participation, even after withdrawal and in addition to regular pay

Voluntary Participation/Withdrawal: Employer does not mind if employee(s) participate, decision to participate is entirely up to employee, subject may withdraw at any time for any reason without affecting job or pay, researcher may stop the study at any time, *and exposure sample might be obtained after withdrawal*

Alternatives: Only alternative is to not participate in study

Questions: Don't sign ICF unless questions are adequately answered, provides contact information for Study Director, Sponsor, and IRB, and *provides a space to indicate Product Name and medical emergency phone number*

Consent: Volunteer signature indicates they read the form (or it was read to the volunteer), *they understood it well,* all questions were answered, they voluntarily agreed to participate, they agreed to release research records, and no legal rights were given up; Researcher signature confirms consent was fully informed and voluntary; Witness signature, if needed, confirms the form was accurately read to the volunteer; *Form also documents if the volunteer wanted to receive their personal study results*

Non-coercion Statement for Employers

An important ethical consideration was that potential volunteers (i.e., employees) must not feel influenced by their employer one way or the other regarding their participation in an AHETF study.

Pre-rule ICFs stated that the prospective employer would not urge, influence or encourage the employee to take part in the study. For post-rule studies, AHETF developed a form that was provided to the employer after having given permission to recruit their employee(s). This form was called an "Employer Cooperation Statement" which provided written assurance that the employee(s) would not suffer any consequence regardless of their decision to participate or not, there would be no coercion of the employees, and that the employees would still receive their normal pay.

Recruitment Meetings with Potential Subjects

Worker recruitment for pre-rule studies was less formal than for post-rule studies. Typically, the CRO identified cooperating employers who then gave permission to recruit employees. Interested workers were identified and the Study Director (SD) discussed pertinent study information with the worker, either one-on-one or in a group.

Post-rule worker recruitment was more formalized. After the employer had signed an Employer Cooperation Statement, the SD was permitted to interact with employees. This was done by sending, or posting on-site, an IRB-approved recruitment flyer (English and/or Spanish) that explained the general aspects of the study, or by conducting an on-site visit. The SD then organized a recruitment meeting with interested workers. Meetings were held one-on-one or with a group of workers. In either case, supervisors or managers were not allowed to attend. At the meetings, the SD discussed AHETF research goals, including eligibility criteria, procedures used in exposure monitoring, and risks/benefits for potential volunteers. A copy of the ICF was also provided and interested employees were encouraged to take a copy home for review (this meeting did not involve signing the consent form).

Informed Consent Meetings with Potential Subjects

The Study Director, or a researcher designated by the SD, was responsible for obtaining consent from a potential subject who expressed interest after attending a recruitment meeting. The consent meeting was held privately with each volunteer. However, the volunteer was given the opportunity to have a friend, family member or adviser attend the meeting. If a volunteer self-identified as being of low or limited literacy, the consent form was read to the volunteer and accommodations were made for a witness to attend. During the meeting, all sections of the ICF were explained in detail and volunteers were encouraged to ask questions. For studies conducted in California, the 'California Experimental Research Subject's Bill of Rights' was also discussed. When the person obtaining consent was satisfied that the volunteer understood the requirements/risks of the study, and the volunteer still wanted to participate, the volunteer was asked to sign and date the ICF and a copy of the signed form was provided to the volunteer. The SD or designee obtaining consent did not sign the ICF unless they believed the worker understood the information presented.

Pre-rule ICFs contained elements required by the IRB as well as some that were specific to the particular study. Consent forms were written in fairly straightforward language with a goal of an 8th grade reading level (for both English and Spanish versions) and were about five pages in length. Although the standard requirements were met in both pre-rule and post-rule ICFs, the post-rule forms were much more comprehensive and detailed (and about twice as long as pre-rule forms).

Assuring Understanding of Consent Form by Potential Subjects

For pre-rule studies, the researcher obtaining consent did not have a formal mechanism to ascertain that the volunteer understood the information presented. However, if a witness was present (because the ICF was read to the volunteer), the witness had to confirm that the volunteer apparently understood the information contained in the consent form.

For post-rule studies, the SD (or designee) did not sign the ICF unless they believed the worker fully understood the information presented. This was ascertained by providing repeated opportunities for the worker to ask questions and/or by asking the workers questions that would require a response indicating understanding of key issues. If the worker was not able to adequately respond to the question, the ICF section was noted, and the section of the ICF where this was discussed was re-reviewed until understanding could be documented. This documentation was maintained in the study records.

Addressing Spanish Speakers

AHETF provided a bilingual researcher for situations in which a worker's preferred language was Spanish. Pertinent documents were provided in Spanish (recruitment flyer, Employer Cooperation Statement, consent form, California Experimental Research Subject's Bill of Rights) and all discussions were conducted in Spanish. When requested by the subject, their personal study results were provided to them in Spanish. During the monitoring process, the bilingual researcher interacted with workers at all times.

Post-exposure Follow-Up

After all MUs within a study were completed, AHETF followed up on two topics. One was to characterize the 'representativeness' of the MUs included in the particular scenario(s). The other was to provide individual exposure-related results to the volunteers within a completed scenario. These activities did not occur with pre-rule studies.

EPA and HSRB requested that AHETF characterize its MUs and demonstrate representativeness of the target population. Individuals knowledgeable with agricultural practices in a particular monitoring area, such as extension agents or university experts, were contacted for their input into how certain pesticide use characteristics related to the MUs compared to all local employers in the same monitoring area (i.e., the local population of employers and workers). Questions such as typical farm size and prevalence of commercial applicators were asked. Determination of representativeness was done for a few monitoring areas and in some scenarios. Information that was collected indicated the equipment and work activities for the MUs collected were generally representative of normal practices for those areas.

Post-rule, study participants were given the opportunity to request a summary of their personal study results. The purpose was to make a worker aware of where on their body pesticide exposure could occur (including a comparison to others who were monitored performing the same activity), why it might have occurred, and general advice on what they could do to minimize dermal exposure. A worker who wanted their results made this request during the informed consent meeting.

Monitoring Area Selection

The identification and selection of monitoring areas for a particular scenario was a key element in designing scenario-specific MU Selection Plans (as described earlier), and plans described in detail the process and basis for selecting monitoring areas for post-rule scenarios. Study protocols then specified exactly how monitoring areas were defined, how many MUs would be collected in each area, and procedures for how monitoring areas could be expanded, if needed to get sufficient cooperators and MUs. Protocols included additional information such as minimum acreages for crop growers, sources of employer lists, and whether lists would include employers without phone numbers (usually they did not).

Recruitment Procedures

After defining a scenario and getting a protocol(s) approved to collect appropriate MUs, the next step in MU collection was to identify and recruit field "cooperators" who could help provide the equipment, location, and worker needed to conduct an MU. Cooperators were owners or managers of farms or commercial application companies or other facilities (such as golf courses or greenhouses) who had the authority to let AHETF pursue volunteers. Once suitable employers were identified, then individual employees were recruited who might volunteer to be a subject and be monitored for exposure. This section describes the various procedures used by AHETF to recruit cooperators and subjects.

Pre-rule, the only mechanism used for finding suitable cooperators or employers was to work with CROs that conducted pesticide research activities, such as efficacy studies, field residue trials, and environmental fate studies. These CROs knew local growers and commercial applicators that were likely to use the pesticide surrogate chemicals and handling techniques that AHETF was researching. AHETF usually contracted with a CRO to act as a "Local Site Coordinator" for each monitoring area and they were responsible for finding suitable cooperators in their area. To facilitate locating cooperating employers, earlier AHETF protocols identified a Principal Field Investigator (PFI) who worked for a CRO and used their own contacts to identify suitable farms or other operations. For multi-area studies, the PFI often utilized CROs in other areas who could more readily find suitable employers for the specific need. In some cases, a local CRO was identified in protocols as a Local Site Coordinator (LSC). From cooperators identified by CROs, subjects were usually chosen by employers or CROs, primarily considering who had appropriate experience and would be available for a scheduled MU. As long as the chosen workers were using appropriate products, equipment, and practices for the specific protocol, they were allowed to participate in the study.

Post-rule studies involved a variety of new procedures designed to ensure equitable selection of subjects and to incorporate "random elements" into the employer and subject selection processes. Employers were often selected from large public lists and subjects were chosen by researchers considering which workers were interested, had appropriate experience, and would be available for a scheduled MU. When multiple employers or workers were available, a random selection was made, such as by flipping a coin.

The first several post-rule studies involved just one new procedure for identifying potential cooperators (which was implemented within each monitoring area). This procedure involved listing employers that might be qualified for the study and randomly identifying a subset(s) of employers for AHETF to contact. These lists mostly came from marketing research companies (for growers) and state regulatory licensing agencies (for commercial pesticide applicators). The initial contact was from a call center that merely asked employers a few questions to determine whether they are "qualified", generally because they use pesticides in a manner consistent with the particular study. Then, an AHETF researcher called all of the qualified employers, introduced the AHETF, explained the study, and asked whether the employer might be willing to cooperate by allowing AHETF to recruit their employee(s). Those employers that expressed interest were termed "Potentially Eligible" and their names were passed on to the SD to confirm eligibility and for scheduling MUs. The goal was to identify many employers and many employees that were "Eligible" and collect the desired number of MUs with protocol-specified diversity in a discrete period of time. Whenever possible, employers and subjects would be selected randomly.

This new process worked fairly well for airblast and aerial application studies which involved fairly common application techniques. The next set of studies involved less common handling tasks and some non-farm settings such as rights-of-way, greenhouses, and nurseries. AHETF sometimes called thousands of employers just to find a single MU, so other techniques for identifying cooperators were implemented, always in conjunction with the list and call techniques just described.

For three studies, protocols specifically allowed employers previously "known to AHETF" to be added directly to the list of qualified growers. In two cases, AHETF had conducted surveys and/or interviews of selected employers during the scenario planning process and it was desirable to include those employers on the qualified list. For example, AHETF had called many growers to identify those that apply liquids by ground followed by immediate incorporation into the soil to determine what products were used and had called many right-of-way commercial applicators to discuss application techniques. Another form of listing known employers was instituted for the closed loading of liquids scenario where the NAAA provided a list of aerial applicators that AHETF suspected were very likely to utilize closed systems for transferring liquids from returnable containers. These employers were deemed to be automatically qualified and so were added directly to the list of qualified employers, thus averting the need for call center screening.

Another technique that was developed for some post-rule studies was to identify "Local Agricultural Specialists" (or LASs) who could identify employers who are expected to qualify for the particular study. Locally knowledgeable agricultural specialists were contacted by AHETF researchers and asked to identify employers that the LAS believed likely to qualify for the particular study. LASs generally worked for contract research organizations or crop consultants. These LASs identified and contacted employers that they believed would be qualified for the particular study, which generally meant they handled pesticides in a certain manner. This contact involved asking the employer to confirm they (or their employees) handled products in that way and obtaining permission to provide their name, address, and phone number to AHETF. This permission was necessary since the employers were not part of a publically-available list. The SD ensured these local specialists were not recruiting potential cooperators, but merely determining whether they were qualified because they (or their employees) handled pesticide products in the appropriate way. Because these referrals were based on personal knowledge and contact from the local specialists, the employers were considered qualified employers. For this reason, it was not necessary that these employers be contacted by a call center and be pre-screened as with large public lists of growers. As with employers known to AHETF discussed earlier, employers identified by LASs were considered to be qualified and were added directly to the list of qualified employers.

Finally, some studies allowed employers or subjects to refer other employers that were likely qualified for a particular study. In most cases, one employer identifying another was not considered suitable since the identified employer was not randomly chosen, and it was believed this could lead to a biased sample. However, two late studies specifically allowed one grower to refer another grower onto employer lists: the open cab ground boom application with soil incorporation study and the granule loading study. In both cases, AHETF found very few qualified growers by the list and call procedures, so a referral from one grower of another was allowed for identifying additional new qualified employers. For the ROW application study, many employers (commercial applicators or utility companies) performed ROW applications in more than one monitoring area (i.e., multiple states), but were not always listed for the other monitoring area(s). By protocol amendment, employers were allowed to be added to other monitoring area lists when they indicated they make ROW applications in that other area. This is another form of employer "referral" that was used for a special case.

Acceptable MU Configurations

As discussed earlier, several statistically equivalent MU configurations were identified, such as five monitoring areas with five MUs per area. Post-rule study protocols specified the desired configuration (based on the MU Plan which presented the statistical justification for sample size and diversity requirements) as well as procedures for handling situations when obtaining the desired configuration was not possible. For example, when employer lists were exhausted and sufficient MUs had not yet been obtained, protocols specified procedures to expand the monitoring area, such as by adding grower names by looking up phone numbers, including other crops, and/or expanding the geographical description of the monitoring area.

The concept of acceptable configurations as described earlier was not considered in pre-rule studies. For pre-rule scenarios for which there were no existing data, protocols were typically designed for five MUs in five monitoring areas, equivalent to 25 MUs for the scenario.

Review of Scenario Plans and Study Protocols

Prior to implementation of all AHETF worker exposure monitoring studies, the scenario plan and study protocol were reviewed by the Joint Regulatory Committee (JRC) made up of North American regulatory authorities, an independent institutional review board, (IRB) and, if required, by the Human Studies Review Board (HSRB) (i.e., post-rule intentional dosing studies). This section describes the process for external review of Monitoring Unit (MU) Plans and study protocols which is described in Table 2. After each review step, AHETF revised the documents to address comments by the reviewer(s) until the reviewer was satisfied, then proceeded to the next step.

Table 2. Steps for Review of MU Plans and Study Protocols

Step	*AHETF Prepared*	*Reviewed By*	*Comments*
1	Draft: MU Plan	JRC (science and ethics)	All post-rule scenarios
2	Revised: MU Plan Draft: Protocol, Consent Form, others[a]	IRB (ethics)	All protocols
3	Revised: MU Plan, Protocol, Consent Form, etc.	California Department of Pesticide Regulation (CDPR) (science and ethics)	All protocols involving testing in California
4	Revised: MU Plan, Protocol, Consent Form, etc.	EPA (science and ethics)	Post-rule Observational studies
		EPA and then HSRB (science and ethics)	Post-rule Intentional Dosing studies
5	Final Draft: MU Plan, Protocol, Consent Form, etc.	IRB (ethics)	All protocols
6	Revised: MU Plan, Protocol, Consent Form, etc.	CDPR (science and ethics)	All protocols involving testing in California
7	Fully approved: MU Plan, Protocol, Consent Form, etc.	Study implemented only after all approvals were received and protocol was signed	

[a] IRB reviewed and approved all Informed Consent Forms, Recruitment Flyers, Product Risk Statements, letters to employers, and any other materials given to workers during recruitment.

Conduct of Exposure Monitoring Studies

The focus of this section is on the study protocol itself since that document defines all the methods used to conduct a particular study. The assigned Study Director was responsible for preparing the study protocol for review. Pre-rule protocols were similar to post-rule protocols, except they did not include the extensive ethical considerations. AHETF developed a standard protocol format that addressed all applicable items, including: rules and regulations; study management practices; procedures for selecting cooperators and field sites; exposure monitoring techniques, field fortification instructions; worker and environmental observations; and archiving procedures.

Management of Studies

The Agricultural Handler Exposure Task Force, LLC was both the Study Sponsor and the Testing facility. Testing facility management consisted of member company representatives (i.e., Sponsor Representatives) serving on various committees and subcommittees.

Testing facility management designated a Study Director who was responsible for the conduct of each study. These responsibilities ranged from assuring that the protocol was approved and followed, that all experimental data were recorded and verified, and that all required items were transferred to archives before the study was completed. In addition, the Study Director was responsible for overseeing the field and analytical phases of each study. For IRB purposes, the Study Director was identified as the "Principal Investigator" for each AHETF study. The role of Study Director is mandated by GLP regulations (40 CFR Part 160.33).

Field Monitors were company representatives who visited a study site for purposes of observing the respective field and analytical activities. Field Monitors provided assistance to the Study Director and oversight of the specific monitoring unit (MU).

Analytical Monitors were consultants contracted by AHETF to perform various functions as directed by the Analytical Subcommittee. Analytical Monitors were responsible for reviewing and evaluating analytical data and kept the Study Director updated on analytical activities, including method validation, sample analyses, and reporting analytical results.

The Quality Assurance Unit (QAU) was an independent AHETF contractor responsible for assuring that applicable GLP requirements were adhered to. The QAU was mandated by GLP regulations under 40 CFR Part 160.35. The responsibilities of the QAU include: reviewing draft protocols and SOPs, auditing field activities to ensure compliance with GLPs, protocols, and SOPs; reviewing raw data; and auditing final field and analytical reports and monographs.

Field Methods and Equipment

The test system for all studies was the worker who handled pesticides according to label directions. Workers were growers or employees of growers; commercial applicators or employees of commercial applicators; or other facility owners or employees of those facilities. Experience performing the particular activity was always a study inclusion criterion.

The protocol described the handling activity being monitored and possible associated mixing/loading and/or application equipment. As appropriate, these descriptions contained specific information about what procedure(s) and equipment were not acceptable. Workers were only allowed to use the type of equipment with which they were familiar, and they were required to have recent experience (protocol-specific) with the equipment and activity.

Protocols specified the use of a field study team that was sufficiently staffed to conduct the required MU-related monitoring activities. At a minimum, those on site during the monitoring of a worker included the following: the Study Director (or designee); an individual who assisted in the dressing/undressing of the worker and collected the exposure samples; when necessary, a female team member who supervised pregnancy testing for female workers and assisted in their dressing/undressing; an observer who took notes on the activities of the worker and recorded specific details such as clothing/PPE and product used; an individual who prepared and conducted field fortifications when required; and a bilingual researcher who offered communication between the worker and study personnel when needed. Post-rule, a medical professional was also present, and often served as the observer, as well.

Test substances were defined as a commercially available registered product used in AHETF studies that contained an active ingredient that had been pre-approved as a surrogate. A sample from each lot of test substance used in a study was collected and sent to a designated laboratory for GLP purity analysis (content of active ingredient in the test substance). GLP purity analysis was performed concurrently with the analytical phase of the study and was reported in a certificate of analysis (COA).

All AHETF protocols specified active ingredient handled (AaiH) as the normalizing factor. The AaiH was calculated from the amount of test substance handled multiplied by the AI content according to the COA (or the nominal concentration if a COA was not available).

Exposure Monitoring Techniques

The passive dermal and inhalation monitoring techniques used in AHETF studies were described earlier in this chapter. For dermal exposure, study protocols indicated: whether dermal dosimeters would be cut in two or six sections, whether or not foot exposure would be monitored, whether or not head patches would be utilized to measure exposure under chemical-resistant hats, and what type of inhalation dosimetry would be used for each possible surrogate. For inhalation exposure, protocols specified the desired flow rate for sampling pumps, specified pumps would be turned on immediately prior to the start of the monitoring period and will operate continuously until the end of the period.; and specified how to handle pumps that stop operating during the work cycle (replace with a pre-calibrated replacement pump or give fresh batteries as soon as possible, and measure the flow rate before monitoring continues). Protocols also specified the following order of sample collection at the end of the monitoring period in order to minimize cross contamination: take inhalation sample(s), remove personal protective equipment (PPE), take outer heat patch sample, take inner head patch sample, take inner sock sample, take hand wash sample, take face/neck wipe sample, and finally take inner dosimeter samples. The monitoring time of an MU was based on pump start and stop times.

Control of Exposure Measurement Bias

Exposure measurement bias was controlled by monitoring workers over a period representative of a typical work day, and by sampling over the entire body of each worker. In addition, quality control samples in the field and laboratory also acted as methods for controlling bias.

Field Fortifications

Field fortifications have always been an integral quality control feature of AHETF monitoring studies since they provide a measure of active ingredient stability during field collection, transportation, and storage. Protocols specified how often fortifications should be conducted and specified the fortifications levels that would be used.

Pre-rule protocols called for fortifications to take place on representative days during the study provided that the location and environmental conditions were similar. Furthermore, there were four levels available for each matrix (since some protocols involved multiple handler tasks that could lead to different exposure levels). On each day that field fortifications occurred, samples of each matrix were fortified with at least two of these levels. The levels chosen were based on expected exposure levels for the handler task being monitored, and the matrix/level combination was prepared and analyzed in triplicate.

Post-rule protocols called for field fortifications to be conducted on a minimum of one day of exposure monitoring for each set of MUs for each active ingredient used in a monitoring area, or more days as appropriate for environmental. Each matrix was fortified at three levels with each level prepared and analyzed in triplicate.

Worker Observations

All AHETF study protocols indicated that worker observations would be taken. For each MU, a researcher was assigned the task of 'observer' and was responsible for documenting any occurrences that might affect worker exposure. Other responsibilities of the observer including: keeping track of the amount of test substance handled, describing the clothing and PPE worn by the subject, and assuring the portable sampling pump was operating normally.

Environmental Monitoring

Study protocols indicated MUs should not be conducted under inappropriate field conditions (e.g., excess rain or wind) and required environmental conditions such as air temperature, relative humidity, wind speed, and wind direction be measured and recorded by means of a calibrated, on-site, portable weather station during all monitoring.

Sample Identification, Storage, and Shipping

Protocols specified that all worker and field fortifications samples be identified by unique sample numbers assigned by AHETF. All samples collected in the field were immediately placed into frozen storage as soon as possible after collection and the time samples were placed in storage was recorded. In addition, the analytical laboratory storage samples were kept under frozen conditions until analysis.

Data and Sample Retention Practices

Study protocols specified an AHETF archive facility and required that all field raw data, test substance samples, laboratory raw data, and original signed study reports be sent to archives. Sample extracts, extracted sample matrices, unanalyzed fortification matrices, and analytical standards were retained in the laboratory until the Study Director and Analytical Monitor determined they were no longer useful. This was generally after the study reports and all associated monographs were submitted to the appropriate regulatory agencies.

Analysis of Exposure Monitoring Samples

AHETF studies were performed using surrogate chemicals as representative examples of all active ingredients that may be used in a given scenario. In order to aid the various analytical laboratories contracted to perform method development, method validation, storage stability, and sample analysis in their work, AHETF prepared a set of analytical guidelines. The guideline document contained directions for handling worker samples, laboratory fortification samples, and field fortification samples; analytical method development and validation procedures; sample quantitation; and reporting of results. The following sections provide more information about these topics and summarize the AHETF analytical guidelines.

Analytical Methods

For every passive dosimetry matrix used to determine residues of surrogate active ingredients, a validated analytical method was used to ensure the accuracy of the measured residues. This means a validated analytical method had to be available for every matrix/surrogate combination that was used in AHETF studies.

During the review of existing studies, AHETF Analytical Monitors reviewed study reports that were being considered for purchase to ensure the analytical results were acceptable. The review consisted of ensuring that validated analytical methods were used; that field fortifications and concurrent laboratory fortifications were conducted at appropriate levels; and that those recoveries met the acceptance criteria.

Analytical methods used by the AHETF originated in one of three ways: 1) they were developed for use by a previous Task Force (e.g., Agricultural Reentry Task Force, ARTF and ORETF), 2) they were provided by an AHETF member company following development for use in a company-sponsored worker exposure study, or 3) they were developed at a contract laboratory sponsored by AHETF specifically for use in AHETF studies.

Analytical Limits of Detection and Quantitation

The Limit of Quantification (LOQ) for all AHETF studies was defined as the lowest level fortified for a matrix or the lowest analytical standard in a study where acceptable recovery was achieved. Target LOQ levels were determined for each analyte/matrix combination based on the analytical method sensitivity and anticipated exposure levels. In general, target LOQ values were 1 μg/sample for inner dosimeters, socks, face/neck wipe, and hand wash samples, and 0.010 μg/sample total for OVS tubes (0.0050 μg/section). The target level was then validated to ensure acceptable recovery and the established LOQ for each matrix was defined in the appropriate analytical method.

Pre-rule, the limit of detection (LOD) was arbitrarily set at 0.3 times the LOQ value. Post-rule, the LOD was defined as the lowest concentration of an analyte that could be determined to be statistically different from a blank. If there were sufficient recoveries for an analyte performed at the LOQ in each matrix ($n \geq 7$) to statistically determine the LOD values based on method validation data, the LOD was set based on EPA Guidance (*17*).

Analytical Method Validation

Following method development, analytical methods were validated for each analytical method/matrix combination. Method validation typically consisted of at least two independent analytical sets, with a minimum of seven samples per set. Each analytical set was comprised of at least one control, three fortifications at the LOQ for the analytical method, and three fortifications at a higher level (typically 10X to 100X the LOQ). Acceptable validation required recoveries at each fortification level between 70% and 120% of the fortification level with a Percent Relative Standard Deviation (%RSD) of ≤20%.

If a laboratory implemented an analytical method for the first time, a method tryout was performed prior to analysis of field-generated samples. If an LOD was not previously determined, the method tryout required a minimum of one control and seven fortifications at the LOQ to allow for calculation of an LOD. If an LOD had been previously determined, the method tryout consisted of a minimum of one control and three fortifications at the LOQ. As with method validation, acceptable recoveries were defined as between 70% and 120% of the fortification level with a %RSD of ≤20%.

Storage Stability Studies

In some cases, storage stability data in the appropriate AHETF matrices was available from member companies. When data were not available, storage stability studies were performed in support of the analytical methods for each surrogate chemical and exposure matrix type used in AHETF studies. These studies were designed to ensure the integrity of worker exposure samples between the time of collection and analysis (i.e., during frozen storage), were based upon the guidelines contained in Series 875, Part A of the U.S. EPA Pesticide Assessment Guidelines: Occupational and Residential Exposure Test Guidelines, and were conducted in accordance with EPA FIFRA Good Laboratory Practice Standards (GLPs).

Samples tested for each surrogate typically consisted of inner dosimeters, face/neck wipes, hand wash, and OVS tube matrices. Cloth head patches and socks were occasionally used as exposure matrices by AHETF; in those instances storage stability results for inner dosimeters were used to support these matrices. Typically, pre-determined amounts of the appropriate surrogate chemical were applied to dozens of samples of each matrix, placed into frozen storage, and triplicate samples were removed periodically to analyze for recovery. Stability studies usually ran for 6 months, but if exposure monitoring study samples were stored for longer periods of time then longer storage stability samples were also obtained.

Storage stability results were reported to support AHETF field and analytical studies and reports were usually submitted to regulatory agencies along with the study report or monograph report that they originally supported.

Analytical Reporting

Analytical reports were prepared for inclusion in all AHETF study reports. Pre-rule AHETF studies usually used a single surrogate compound making analytical reporting relatively straightforward. A single laboratory was assigned responsibility for analytical support for a given study and handled method development and validation, field fortification preparation, and all sample analysis for that study.

Post-rule studies often included multiple surrogate chemicals that could be used in a single study, with analyses potentially occurring at multiple laboratories. In response to these changing requirements, AHETF analytical assignments were changed to assign each surrogate chemical to a particular analytical laboratory. Each laboratory would provide analytical support for its assigned surrogate chemicals across all studies utilizing each surrogate. At the completion of a study, the analytical results and supporting data for each surrogate chemical were presented in an individual analytical report. The individual analytical reports for each surrogate chemical, often prepared by multiple analytical laboratories, were then included in the final study reports as appendices.

As mentioned earlier, method validation reports and storage stability reports were also prepared. These were typically submitted to regulatory agencies along with the first study report or monograph report that they supported. Subsequent reports and monographs utilizing those surrogates and exposure matrices then cited the submitted analytical reports.

Reporting Field Study and Scenario Results

All AHETF exposure studies were conducted under GLP regulations; accordingly, one "study report" was produced for each field study protocol. Pre-rule, study reports included all supporting information for study conduct plus results for every monitoring unit (MU). This included residues adjusted for field fortification and normalized total dermal exposure (nTDE) and normalized total inhalation exposure (nTIE). Post-rule, two important changes were made: 1) post-rule AHETF study reports did not include results adjusted for field fortifications, and 2) a new separate report was prepared that presented all correspondence between AHETF and the IRB that approved the study protocol.

Also, post-rule, AHETF began submitting scenario monographs which provided calculations for nTDE and nTIE. AHETF generated scenario monographs for the benefit of regulators and other potential database users. Each monograph includes a description of the scenario, a summary of all the MUs representing the scenario, and an assessment of the data adequacy for that scenario. The following sections briefly describe the content of AHETF monograph reports.

Exposure Calculations

Because study reports evolved to include only unadjusted residues found in field samples, the monographs presented details for all of the calculations performed - from raw, unadjusted residues to nTDE and nTIE calculations for every MU in the monograph. It is important to note that existing study MUs were handled exactly as AHETF-conducted MUs in terms of these exposure calculations. The following sections describe important calculation techniques used by AHETF.

Adjusting for Field Fortifications

AHETF has always adjusted all quantifiable individual worker exposure matrix results for appropriate field fortification (FF) results (for both purchased and AHETF-conducted studies). The exact process was fully described in scenario monographs, but the following conventions were followed for each exposure sampling matrix used:

- Adjustment factors were calculated for each level of fortifications (typically three levels)
- Adjustments were based on the level that is closest to each individual residue to be adjusted (based on mid-points between successive levels as the cutoff)
- Adjustments were made for all samples, upward or downward, except adjustment factors were capped at 1.20 (reflecting 120% average recovery)

Because more complex recruiting procedures were implemented post-rule, AHETF generally shifted from using study-wide average fortification results to using MU-specific average fortification results for most post-rule AHETF studies. This was primarily because MUs became so spread out in time that many more fortification events were necessary, but it also reflected the fact that post-rule studies often involved the potential for using several surrogate chemicals per scenario (which again resulted in more fortification events).

It should be noted that EPA has directly used the FF-adjusted exposure results produced by AHETF, but other regulatory agencies sometimes re-calculate adjusted residues according to their specific procedures.

Adjusting for Face PPE Worn

Post-rule, AHETF extrapolated face/neck exposure results upward for subjects that wore protective eyewear and/or respiratory protection devices during monitoring. This extrapolation was needed to account for the fact that some face and/or neck area that would otherwise be exposed and wiped was covered by the PPE. This extrapolation was done by multiplying the face/neck residue for a subject by an appropriate PPE-extrapolation factor based on estimates of surface areas covered by various face PPE. This resulted in face/neck residue measurements that reflect no special PPE worn by workers. All AHETF monographs include such adjustments for purchased and AHETF-conducted MUs.

Extrapolating Face/Neck Exposure to Whole Head Exposure

AHETF has always extrapolated face/neck exposures to the rest of the head that is normally covered by hair when face/neck wipe residues were the only head dosimetry used (i.e., when there were no hat or head/body patch dosimeters that could represent head or neck parts). This extrapolation was based on relative surface areas of the face/neck area that was wiped and the rest of the head that was not wiped. The surface areas are gender-specific and are based on EPA values from their Series 875 guidelines and standard reference values identified by the North American Free Trade Agreement (*18*). The only exception is a monograph for the On-Farm Seed Treatment with Solids (OFST-S) scenario that did not extrapolate face/neck exposure to the rest of the head.

Calculating Total Inhalation Exposure

As described in 5.1.5, inhalation monitoring involved a personal air sampling pump and a collection device attached to the worker's collar and facing downward. Air was drawn through the device at about 2 L/min and flow rates were measured at the start of monitoring and at the end of monitoring. A daily inhalation exposure value was estimated for the worker based on an assumed breathing rate as follows:

Inhalation Exposure =
Total Residue (μg) X Breathing Rate (L/min) / Flow Rate (L/min)

The assumed breathing rate is dependent on the particular task being performed and AHETF used the following rates (*18*):

- 8.3 L/min for sitting activities such as piloting an aircraft and driving a tractor
- 16.7 L/min for light activities such as mixing/loading light packages (< 50 lb) and applications while standing or walking; and
- 26.7 L/min for moderate activities such as mixing/loading heavy packages (> 50 lb) and backpack applications

In addition, a breathing rate of 9.1 L/min was used for the OFST-S scenario which reflects a time-weighted average breathing rate accounting for some time treating seed (at 16.7 L/min) and more time planting treated seed (at 8.3 L/min).

It should be noted that regulatory agencies might use different assumptions for breathing rates to estimate total inhalation exposure.

Normalizing Total Exposure

Normalized worker exposure is obtained by dividing the total exposure for the monitoring period by the worker's NF. Both total dermal and total inhalation exposures are normalized in this manner. All monograph reports presented normalized exposures for all MUs and used them when calculating statistics and performing statistical analyses related to accuracy.

Scenario Statistical Analyses

Because no scenario monographs were submitted pre-rule, all of the AHETF monographs include statistical analyses that reflect guidance provided in the Governing Document (assuming there were sufficient data for such analyses). However, many scenarios required some modifications to the procedure to account for special structure and potential correlation among MUs. Special structure was often present when existing studies contributed some of the MUs and when MUs were collected in a different configuration than called for in the study design. In addition, individual scenario monographs can be consulted for details about how the analyses were conducted. In brief, the following items were addressed in scenario monographs for both dermal and inhalation exposures:

- Normalized total exposure was calculated for each MU
- Common statistical measures that characterize the distribution of normalized exposures were calculated, including the geometric mean (GM), the arithmetic mean (AM), and the 95th percentile (P95)
- Geometric standard deviation (GSD) and Intra-cluster correlation (ICC) were calculated for the data set based on the structure of the data (i.e., clustering of MUs)
- Statistical measures were calculated based on empirical methods, assuming a lognormal distribution, and based on simulations that took into account the structure of the data
- The primary benchmark objective was evaluated by comparing a calculated 95th percentile accuracy value to the objective (typically 3-fold accuracy)
- Examine the impact of ignoring clusters on statistical measures (i.e., whether simple statistics are similar to model-based statistics)
- The secondary benchmark objective was evaluated by regressing log of total exposure against log of the NF and examining the width of the confidence interval (an interval less than 1.4 indicates 80% power to distinguish proportionality from independence)

For two scenarios, AHETF purchased a single study and prepared a monograph based on that one study that represented just one or two clusters (the closed loading of granules scenario and the backpack MLA of liquids scenario). Because these datasets had few MUs and limited diversity among MUs (e.g., test location and date of exposure monitoring), they were not amenable to the complex benchmark statistical analyses AHETF conducted for most completed exposure scenario datasets. That is, model-based statistical analyses based on multiple independent clusters of correlated MUs could not be conducted. Instead, only simple descriptive statistics were calculated and presented. Specifically, the following statistics were evaluated:

- GM, empirical estimates (with no assumption about the underlying distribution of the data)
- AM, empirical and lognormal estimates (assuming a lognormal distribution of the data)
- P95, empirical and lognormal estimates (assuming a lognormal distribution of the data)

Similarly, the secondary benchmark power analysis of the exposure-AaiH relationship was not appropriate when there was only a single cluster (as well as a very small range of AaiH). Nevertheless, AaiH was used to normalize exposures. This was considered a useful and appropriate way to present exposure potential for this scenario since it allows regulatory agencies to estimate exposure for other products and uses.

Glossary of Terms and Abbreviations

AaiH – Amount of Active Ingredient Handled; the normalizing factor for almost all AHETF scenarios

AHED® – Agricultural Handler Exposure Database; contains most of the MU-specific data that were purchased or conducted by AHETF

AHETF – Agricultural Handler Exposure Task Force, LLC

AI, ai, or a.i. – Active Ingredient; the chemical component quantified in worker exposure samples (parent molecule or otherwise)

AM – Arithmetic Mean

ARTF – Agricultural Reentry Task Force; a pesticide industry task force that generated exposure data for workers reentering pesticide-treated areas

BP-ROW – Scenario acronym for Backpack Application of Liquid Sprays in Utilities Rights-Of-Way

CDPR – California Department of Pesticide Regulation

CFR – Code of Federal Regulations; codification of rules and regulations published in the Federal Register by the executive departments and agencies of the federal government of the United States

Cluster – a group of MU(s) that were collected nearby in terms of time and geography that are evaluated for possible correlation of results during the statistical analysis of exposure results for a completed scenario

COA – Certificate of Analysis; a GLP-compliant analysis of the purity of a test substance sampled from an MU, used to calculate AaiH for that MU

CR – Chemical-resistant (e.g., CR gloves)

CSLL-R – Scenario acronym for Closed System Loading of Liquids from Returnable Containers; eventually combined with CSLL-NR data to become the MTL scenario

CSLL-NR – Scenario acronym for Closed System Loading of Liquids from Non-Returnable Containers; eventually combined with CSLL-NR data to become the MTL scenario

Diversity Selection or Purposive Diversity Selection – a method of choosing specific MU conditions that results in some diversity in important study conditions defined in study plans and/or protocols; stratified diversity selection is a practical form of diversity selection that first partitions all possible selection units into non-overlapping groups called strata and restricts MUs within monitoring areas to certain strata

Employer – an independent person (e.g., a farmer) or company (e.g., commercial applicator) that is recruited by AHETF to see if they might provide one or more volunteers to participate in an AHETF study

EPA – U.S. Environmental Protection Agency; an agency of the United States federal government whose mission is to protect human and environmental health (including regulation of pesticides)

FF – Field Fortifications; worker samples that are fortified in the field with active ingredient, then are weathered, shipped, stored, and analyzed along with worker sample to determine stability of residues in/or exposure matrices

FFDCA – Federal Food, Drug, and Cosmetic Act; a set of laws giving authority to the U.S. Food and Drug Administration (FDA) to oversee the safety of food, drugs, and cosmetics

FIFRA – Federal Insecticide, Fungicide, and Rodenticide Act; a federal law that set up the basic U.S. system of pesticide regulation to protect applicators, consumers, and the environment fRA – fold-relative accuracy; a measure of how close an estimated value of an exposure distribution parameter (e.g. GM, AM, P95) is to its true value. It is defined as either (estimated value)/(true value) or (true value)/(estimated value), whichever is larger. A theoretical distribution of possible fRA values can be determined if an exposure distribution and sample sizes are assumed. However, fRA itself can never be calculated from actual data since true parameter values are never known in practice.

fRA95 – an upper 95% confidence limit for the true (but unknown) fold-relative accuracy for certain statistical parameters (AM, GM, P95) of the scenario-wide exposure distribution. Although the actual fRA is unknown (because true values of parameters are unknown) there is 95% confidence that the fRA is no greater than fRA95.

GH/N – Scenario acronym for Mixing/Loading/Application using Powered Handgun Equipment in Managed Horticultural Facilities (i.e., greenhouses and nurseries)

GLP or GLPs or GLPS – Good Laboratory Practice Standards; a set of federal regulations (40 CFR Part 160) that prescribes practices intended to ensure the quality and integrity of data submitted to support pesticide registrations under FIFRA and FFDCA

GM – Geometric Mean

GSD – Geometric Standard Deviation

Handler – A person who handles pesticides as part of his occupation (for example, a farmer or a commercial applicator)

HD – Handler-Day (the set of conditions associated with a particular worker and the scenario-related activities that he performs during a single workday)

HG-ROW – Scenario acronym for Handgun Application of Liquid Sprays in Utilities Rights-Of-Way

HSRB – Human Studies Review Board; established by the Human Studies Rule in 2006 to advise EPA about third-party research involving intentional dosing of pesticides

Human Studies Rule – federal regulation setting requirements for EPA review of certain intentional dosing studies, including establishing the HSRB to advise EPA; 40 CFR Part 26, published February 6, 2006 and effective April 7, 2006; revisions to the rule were published on February 14, 2013 and effective April 15, 2013

ICC – Intra-Cluster Correlation (or, Intra-Class Correlation for clusters); a measure of the correlation (i.e., similarity) between exposure measurements for MUs within the same cluster of MUs

ICF – Informed Consent Form

IRB – Institutional review board; a general term for a committee established to review and approve research involving human subjects

IIRB – Independent Investigational Review Board; the name of the second IRB AHETF utilized during this testing program (which eventually became Schulman Associates IRB, post-rule only)

JRC – Joint Regulatory Committee; representatives from EPA, CDPR, PMRA, and USDA who provided input to AHETF on science and policy matters

LOD – Limit of Detection; the lowest mass of chemical that is detectible for an analytical method for a particular surrogate/matrix combination

LOQ – Limit of Quantitation or Limit of Quantification; the lowest mass of chemical that is quantifiable for an analytical method for a particular surrogate/matrix combination

L/min – Liters per Minute, also LPM; applicable to air sampling pump flow rates and assumed breathing rates used to estimate inhalation exposure for an MU

MOE – Margin of exposure; the ratio of a chemical's no-observed-adverse-effect level to its theoretical, predicted, or estimated dose of human intake

Mixed (or mixed-effects) Model – a statistical model that uses both random and non-random (or 'fixed') terms to describe observed data. The fixed terms are used to represent the mean and the random terms describe both the variation and correlation(s) in the data. A variance component model is a special type of mixed model that uses the same mean term for every observation. If a statistical model has only a single random variation (i.e. 'error') term, but no random correlation term(s), it is referred to as fixed-effects model instead.

ML or M/L – Mixer/Loader or mixing/loading

MLA or M/L/A – Mixer/Loader/Applicator or mixing/loading/application

Monitoring Area or Monitoring Site – equivalent terms used at various times in MU Plans and protocols to describe a discreet region of North America where study MUs would be collected, ranging in size from a few counties to entire states, to EPA Regions, to several adjacent states (however, some later documents used "monitoring site" to refer to the specific farm or land where pesticide handling took place for an MU)

Monograph – an AHETF report that summarizes and evaluates a complete set of exposure measurements for a particular pesticide handling scenario

MTL – Scenario acronym for Mechanical Transfer of Liquids; a combination of CSLL-R and CSLL-NR data

MU – Monitoring unit; refers to a participant in an exposure study, all monitoring conditions, and the resulting exposure information from that participant for one surrogate chemical

MU Plan – a post-rule document prepared to establish a plan to collect new MUs to complete a handler scenario; also referred to as a Monitoring Unit Selection and Construction Plan or Monitoring Unit Selection Plan

NASS – National Agricultural Statistics Service, a unit of USDA that conducts monthly and annual surveys for estimates of production, supply, prices, and other information necessary to maintain orderly agricultural operations; AHETF used NASS data to plan some scenarios and studies, including: identifying areas where certain crop types are grown, identifying areas where certain surrogate chemicals or pesticide products are used, and obtaining estimates of total populations of growers for certain crops/regions

NF – Normalizing Factor; an experimentally measureable quantity that is expected to be proportional to the potential contact the worker has with active ingredient, and is used to normalize final exposure results (this is almost always AaiH)

nTDE – Normalized Total Dermal Exposure; MU-specific TDE divided by the MU-specific normalizing factor, nTDE is generally expressed in µg/lb ai

nTIE – Normalized Total Inhalation Exposure; MU-specific TIE divided by the MU-specific normalizing factor, nTIE is generally expressed in µg/lb ai

OFST-S – Scenario acronym for On-Farm Seed Treatment with Solids

ORETF – Outdoor Residential Exposure Task Force; a pesticide industry task force that generated exposure data for workers making pesticide applications to residential turf and for people reentering pesticide-treated turf areas

OVS – OSHA Versatile Sampler; the filter and sorbent system used to collect airborne residues for almost all inhalation samples

P95 – 95th percentile

PaiC – Potential AI Contact; the amount of active ingredient that a worker is expected to encounter during a workday

PHED – Pesticide Handler Exposure Database; the database with serious limitations that AHETF data are designed to replace

PMRA – Pest Management Regulatory Agency (of Canada); an agency of Health Canada that includes implementation of pesticide laws

PPE – Personal protective equipment; gloves, eyewear, respirators, or other garments or equipment worn by workers and designed to protect the wearer's body from contact with pesticides

PR Notice – Pesticide Registration Notice; notices issued by the Office of Pesticide Programs to inform pesticide registrants and other interested persons about important policies, procedures and regulatory decisions

PRS – Product Risk Statement; a document that describes the risks of handling a specific pesticide product likely to be used by a subject in an AHETF study (prepared by AHETF based on the product label and safety data sheets)

QAU – Quality Assurance Unit

RSD – Relative Standard Deviation

SAP – Scientific Advisory Panel

Scenario – a well-defined worker exposure situation, usually characterized by specific agricultural chemical handling tasks and equipment (e.g., open pour mixing/loading of liquid products)

Similarity Restriction – a protocol-specified "rule" that limits how similar MUs within a monitoring area can be; they are sometimes mandatory and sometimes only desired

SOP – Standard Operating Procedure; one of an extensive set of written procedures that guided AHETF research (as required by GLPs)

SRF – Study Review Form; a brief summary of an existing study prepared by a member company or AHETF that documents decisions for acceptance or rejection by JRC

SRR – Secondary Review Report; a detailed summary report prepared by AHETF for an existing study for use in JRC review for acceptance

Subject – a person who volunteers to be in an AHETF study and actually participates by starting a monitoring unit (which officially begins when long underwear are put on)

TDE – Total Dermal Exposure; MU-specific sum of the adjusted residues for the inner whole-body dosimeter, total head/neck, total hand wash, and foot, if measured (expressed in µg)

TIE – Total Inhalation Exposure; calculated exposure for an entire monitoring period based on adjusted residues on inhalation sampling devices and assuming a breathing rate for the worker (expressed in µg)

USDA – United States Department of Agriculture; a department of the United States government that manages various programs related to food, agriculture, natural resources, rural development and nutrition (and a member of the JRC)

WBD – Whole-body Dosimeter; a set of clothing that covers the entire body of a person (except head, hands, and feet) and is collected after a work period in order to be analyzed for pesticide residue content

WPS – Worker Protection Standard; an EPA regulation designed to protect agricultural workers, including requirements for PPE, restricted-entry times following pesticide applications, safety training, and other elements to prevent poisoning and injury from pesticides

References

1. EPA. PHED SURROGATE EXPOSURE GUIDE - Estimates of Worker Exposure from The Pesticide Handler Exposure Database Version 1.1, August 1998.
2. Stasikowski, M. (EPA). Memorandum to Daniel P. Fay Concerning Revisions to PHED, March 16, 2001.
3. EPA. Protections for Subjects in Human Research, Federal Register: February 6, 2006, Volume 71, Number 24, Pages 6137–6176, 2006. https://www.govinfo.gov/content/pkg/FR-2006-02-06/html/06-1045.htm (Accessed March 29, 2021).
4. EPA. United States Environmental Protection Agency – History and Focus of the Human Studies Review Board. https://www.epa.gov/osa/history-and-focus-human-studies-review-board (Accessed March 29, 2021).
5. EPA, Health Canada, California Environmental Protection Agency. Review of Worker Exposure Assessment Methods – Presented Jointly to the FIFRA Scientific Advisory Panel, January 9–12, 2007.
6. EPA. SAP Minutes No. 2007-03, A Set of Scientific Issues Being Considered by the Environmental Protection Agency Regarding: Review of Worker Exposure Assessment Methods, FIFRA Scientific Advisory Panel Meeting, January 9–12, 2007.
7. Philpott, S. October 27–28, 2010 EPA Human Studies Review Board Meeting Report, Final Report, December 13, 2010.
8. Fisher, C. June 27–30, 2006 EPA Human Studies Review Board Meeting Report, Final Report, October 6, 2006.
9. Brimijoin, W. S. June 27–29, 2007 EPA Human Studies Review Board Meeting Report, Proposed Final Draft V 1, September 27, 2007.
10. Fisher, C. June 24–25, 2008 EPA Human Studies Review Board Meeting Report, Final Report, November 14, 2008.
11. Fisher, C. October 21–22, 2008 EPA Human Studies Review Board Meeting Report, Final Report, December 30, 2008.
12. Fisher, C. Minutes of the United States Environmental Protection Agency (EPA) Human Studies Review Board (HSRB) April 18–20, 2007; May 7, 2007.
13. EPA. Pesticide Registration (PR) Notice 2007-3: The Agricultural Handlers Exposure Task Force, L.L.C., September 20, 2007.
14. EPA. Draft Framework for Developing Best Practices for Recruiting, Screening, Informing, and Consenting Human Subjects of Occupational Exposure Studies with Pesticides, March 15, 2007.
15. EPA. Occupational and Residential Exposure Text Guidelines, OPPTS 875; U.S. EPA, Washington, D.C., 1996.
16. Fisher, C. June 27–29, 2007 EPA Human Studies Review Board Meeting Report, Final Report, December 18, 2007.
17. EPA. Assigning Values to Non-Detected/Non-Quantified Pesticide Residues in Human Health Food Exposure Assessments, EPA Docket #OPP-00570A, March 23, 2000.
18. NAFTA. Dept. of Pesticide Regulation (DPR), California EPA, HSM-98014, April 24, 1998.

Editors' Biographies

David E. Barnekow

David E. Barnekow (Ph.D. in Natural Product Organic Chemistry from Montana State University) is a Technical Laureate at Corteva Agriscience in Indianapolis, IN. During his time at Corteva, Dave has done research in metabolism, environmental fate, residue analysis, field volatility trials, and human exposure studies. Dave has served on the AGRO Division Executive Committee and has co-organized symposia for the ACS. He is the Administrative Chair for the Agricultural Handler Exposure Task Force.

Michael E. Krolski

Michael E. Krolski (Ph.D. in Synthetic Organic Chemistry from Iowa State University) is a Senior Principal Scientist at Bayer CropScience in Chesterfield, MO, where he has done research in metabolism, method development, residue analysis, crop field trials, and human exposure studies. Mike has served on the AGRO Division Executive Committee and has organized symposia for the ACS, the Pan-Pacific Conference, and the 2014 IUPAC Congress. He is the Technical Chair for the Agricultural Handler Exposure Task Force and the Agricultural Reentry Task Force, and is on the technical committee for the Council for the Advancement of Pyrethroid Human Health Risk Assessment.

Indexes

Author Index

Subject Index

P

S

T

U

Printed in the USA/Agawam, MA
March 16, 2022

790247.001